Esprit Basic Research Series

Edited in cooperation with
the European Commission, DG III

Editors: P. Aigrain F. Aldana H. G. Danielmeyer
O. Faugeras H. Gallaire R. A. Kowalski J. M. Lehn
G. Levi G. Metakides B. Oakley J. Rasmussen J. Tribolet
D. Tsichritzis R. Van Overstraeten G. Wrixon

Springer
Berlin
Heidelberg
New York
Barcelona
Budapest
Hong Kong
London
Milan
Paris
Santa Clara
Singapore
Tokyo

J. F. McGilp   D. Weaire
C. H. Patterson (Eds.)

Epioptics

Linear and Nonlinear
Optical Spectroscopy
of Surfaces and Interfaces

With 120 Figures and 6 Tables

Springer

Volume Editors

John F. McGilp
Denis Weaire
Charles H. Patterson

University of Dublin
Department of Physics
Trinity College
Dublin 2, Ireland

Library of Congress Cataloging-in-Publication data applied for

Die deutsche Bibliothek – CIP-Einheitsaufnahme
Epioptics : linear and nonlinear optical spectroscopy of surfaces and interfaces ; with 6 tables /
J. F. McGilp ... (ed.). – Berlin ; Heidelberg ; New York ; Barcelona ; Budapest ; Hong Kong ;
London ; Milan ; Paris ; Santa Clara ; Singapore ; Tokyo : Springer, 1995
(ESPRIT basic research series)
ISBN-13:978-3-642-79822-1   e-ISBN-13:978-3-642-79820-7
DOI: 10.1007/978-3-642-79820-7

NE: MacGilp, John F. [Hrsg.]

ISBN-13:978-3-642-79822-1

Publication No. EUR 16619 EN of the European Commission, Dissemination of Scientific and Technical Knowledge Unit, Directorate-General Information Technologies and Industries, and Telecommunications, Luxembourg. Neither the European Commission nor any person acting on behalf of the Commission is responsible for the use which might be made of the following information.

Typesetting: Camera-ready by authors/editors
SPIN 10084543   45/3144 - 5 4 3 2 1 0 - Printed on acid-free paper

Preface

The study of condensed matter using optical techniques, where photons act as both probe and signal, has a long history. It is only recently, however, that the extraction of surface and interface information, with submonolayer resolution, has been shown to be possible using optical techniques (where "optical" applies to electromagnetic radiation in and around the visible region of the spectrum). This book describes these "epioptic" techniques, which have now been quite widely applied to semiconductor surfaces and interfaces. Particular emphasis in the book is placed on recent studies of submonolayer growth on well-characterised semiconductor surfaces, many of which have arisen from CEC DGXIII ESPRIT Basic Research Action No. 3177 "EPIOPTIC", and CEU DGIII ESPRIT Basic Research Action No. 6878 "EASI". Techniques using other areas of the spectrum such as the infra-red region (IR spectroscopy, in its various surface configurations), and the x-ray region (surface x-ray diffraction, x-ray standing wave), are omitted. The optical techniques described use simple lamp or small laser sources and are thus, in principle, easily accessible.

Epioptic probes can provide new information on solid-gas, solid-liquid and liquid-liquid interfaces. They are particularly suited to growth monitoring. Emerging process technologies for fabricating submicron and nanoscale semiconductor devices and novel multilayer materials, whether based on silicon or compound semiconductors, all require extremely precise control of growth at surfaces. *In situ*, non-destructive, real-time monitoring and characterisation of surfaces under growth conditions is needed for further progress. Both atomic scale resolution, and non-destructive characterisation of buried structures, are required. This has now been demonstrated, using epioptic techniques, with model systems and, most recently, in growth reactors under normal operating conditions of temperature and pressure. Prospects are excellent for the further development of a variety of epioptic probes for semiconductor growth monitoring and characterisation.

John McGilp
Denis Weaire
Charles Patterson

Contents

Contributors

Andrea d'Andrea	*I.M.A.I., C.N.R., Roma*
Michele Cini	*Dipartimento di Fisica, II Università di Roma 'Tor Vergata'*
Rodolfo Del Sole	*Dipartimento di Fisica, II Università di Roma 'Tor Vergata'*
Raffaello Girlanda	*Istituto di Struttura della Materia, Università di Messina*
Jiang Guo-Ping	*Dipartimento di Fisica, II Università di Roma 'Tor Vergata'*
David Hobbs	*Department of Physics, Trinity College Dublin*
John McGilp	*Department of Physics, Trinity College Dublin*
Charles Patterson	*Department of Physics, Trinity College Dublin*
Edoardo Piparo	*Istituto di Struttura della Materia, Università di Messina*
Lucia Reining	*Centre Européen de Calcul Atomique et Moleculaire, Université Paris Sud*
Wolfgang Richter	*Institut für Festkörperphysik, Technische Universität Berlin*
Uwe Rossow	*Institut für Festkörperphysik, Technische Universität Berlin*
Anatolii Shkrebtii	*Dipartimento di Fisica, II Università di Roma 'Tor Vergata'*
Zbig Sobiesierski	*Department of Physics and Astronomy, University of Wales College of Cardiff*
Claudio Verdozzi	*IRC in Surface Science, University of Liverpool*
Denis Weaire	*Department of Physics, Trinity College Dublin*
Dietrich Zahn	*Technische Universität Chemnitz-Zwickau*

Abbreviations

AES	Auger electron spectroscopy
ALE	Atomic layer epitaxy
BZ	Brillouin zone
CCD	Charge-coupled device
DAS	Dimer-adatom-stacking fault
DELS	Diffuse elastic light-scattering
DF	Density functional
DOR	Dynamic optical reflectivity
ECLS	Epitaxial continued-layer structure
EFIRS	Electric-field induced Raman scattering
EM	Electromagnetic
EXAFS	Extended x-ray absorption fine structure
FTIR	Fourier transform infra-red spectroscopy
FWHM	Full width at half maximum height
GVB	Generalised valence bond
IR	Infra-red
JDOS	Joint density of states
LDA	Local density approximation
LEED	Low energy electron diffraction
LLS	Laser light scattering
LMTO	Linear combination of muffin-tin orbitals
LO	Longitudinal optical (phonon)
MBE	Molecular beam epitaxy
MEE	Migration-enhanced epitaxy
ML	Monolayer
MOMBE	Metal-organic molecular beam epitaxy
MOS	Metal-oxide-semiconductor
MOVPE	Metal-organic vapour phase epitaxy
MQW	Multi-quantum well
OMA	Optical multichannel analyser
OPO	Optical parametric oscillator
PDE	Probability density envelope
PEM	Photoelastic modulator

PL	Photoluminescence spectroscopy
PLE	Photoluminescence excitation spectroscopy
PP	Perfect-pairing
QW	Quantum well
RAS	Reflectance anisotropy spectroscopy
RD	Reflection difference
RDS	Reflection difference spectroscopy
RHEED	Reflection high-energy electron diffraction
RPA	Random phase approximation
RRS	Resonant Raman scattering
RS	Raman spectroscopy
RT	Room temperature
SBZ	Surface Brillouin zone
SDA	Surface dielectric anisotropy
SDR	Surface differential reflectance
SE	Spectroscopic ellipsometry
SFG	Optical sum-frequency generation
SHG	Optical second-harmonic generation
SIOA	Surface-induced optical anisotropy
S/N	Signal-to-noise
SO	Strong-orthogonality
SPA	Surface photoabsorption
SRS	Surface Reflection Spectroscopy
STM	Scanning tunneling microscopy
TA	Transverse acoustic (phonon)
TEG	Triethylgallium
TEM	Transmission electron microscopy
TMG	Trimethylgallium
TO	Transverse optical (phonon)
UHV	Ultra-high vacuum
UV	Ultra-violet
VASE	Variable angle spectroscopic ellipsometry
VPE	Vapour phase epitaxy
XPS	X-ray photoelectron spectroscopy

Chapter 1. Introduction

John McGilp

Department of Physics, Trinity College, Dublin 2, Ireland

1.1 Phenomenology

The penetration depth of optical radiation into condensed matter is large, in general, which makes the isolation of a surface or interface contribution difficult ("interface", from now on, is understood to include the surface, which is the condensed matter-ambient interface). Even in the near ultra-violet (UV) region, the interface contribution to the linear reflectivity from a semiconductor will only be a few percent. However, deeper understanding of the underlying physics of the optical response, combined with advances in instrumentation, have allowed the contribution from the interface to be identified. Particularly important has been the recognition that symmetry differences between the bulk and interface can be exploited, as can interface electronic and vibrational resonances [1]. The special efforts and techniques required to obtain this interface optical response, together with the increasing activity in the area, justify distinguishing it from more conventional optical studies. The term "epioptics" has been coined for this special area (from the Greek "*epi*" meaning "upon") [2].

The study of interfaces using epioptic techniques (Fig. 1.1) offers several significant advantages over conventional surface spectroscopies. The material damage and contamination associated with charged particle beams is eliminated; all pressure ranges are accessible; insulators can be studied without the problem of charging effects; buried interfaces are accessible owing to the large penetration depth of the optical radiation. In addition, epioptic techniques offer micron lateral resolution and femtosecond temporal resolution (the latter has hardly been exploited). Non-destructive, *in situ* characterisation of material thin films, surfaces and interfaces in all pressure régimes is central to the development of new materials and processes, particularly in this evolving era of nanoscale structures.

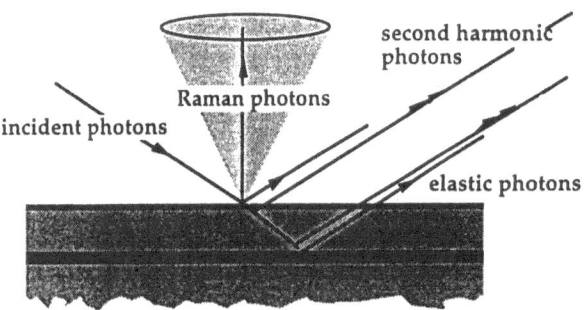

Fig. 1.1. Epioptics

The potential of these optical techniques for thin film characterisation, down to single atomic layers, began to be recognised in the 1980s. A comprehensive review of the application of optical techniques to organic thin films from the submonolayer to micron thickness range appeared in 1987 [3], and a short review of *in situ* epioptics applied to well-characterised systems under ultra-high vacuum (UHV) conditions appeared in 1990 [1].

1.2 Optical Response of Materials

The interaction of an electromagnetic (EM) field of optical frequency, ω, and wave vector, $\mathbf{k}$, with condensed matter, is described by a general polarisation amplitude, $\mathbf{P}(\mathbf{k}, \omega, 2\omega, \dots)$, induced by the field of amplitude $\mathbf{E}(\mathbf{k}, \omega)$:

$$\mathbf{P}(\mathbf{k}, \omega, 2\omega, ..) = \varepsilon_0 \left[\chi^{(1)}(\mathbf{k}, \omega) \cdot \mathbf{E}(\mathbf{k}, \omega) + \chi^{(2)}(\mathbf{k}, \omega, 2\omega) \colon \mathbf{E}^2(\mathbf{k}, \omega) + \dots \right] \quad (1.1)$$

where $\chi^{(i)}$ is the ith-order dielectric susceptibility tensor describing the material response. In general, the susceptibility tensor will show both frequency and spatial dispersion [4], and this will be implicit from now on. In an anisotropic, *linear* medium, the more familiar dielectric tensor, ε, is related to the susceptibility by

$$\varepsilon = \varepsilon_0 \left(1 + \chi^{(1)} \right) \quad (1.2)$$

where **1** is the unit tensor. The other, *nonlinear* terms in equation 1.1 may become significant at high EM field strengths.

In order to show the relationship between the various epioptic techniques discussed here, equation 2.1 is simplified by assuming only scalar variables, and by re-writing it so as to include expansion in a material deformation parameter, Q, and a static field parameter, $E(0)$, as well as the EM field, $E(\omega)$ [5]:

$$P(\omega,2\omega,..) = \varepsilon_0 \{ \chi^{(1)}(\omega)E(\omega) + \left[\partial\chi^{(1)}(\omega) / \partial Q \right] QE(\omega)$$

$$+ \left[\partial^2\chi^{(1)}(\omega) / \partial Q \partial E(0) \right] QE(0)E(\omega) + \chi^{(2)}(\omega,2\omega)E^2(\omega) + ... \}$$

$$(1.3)$$

Table 1.1 lists the various commonly-used epioptic techniques, together with their dependence on susceptibility. The terminology used in this book is shown in **bold**. The evolution of the field has resulted in a confusion of names, and it is hoped that this terminology may become generally accepted. The techniques are described in more detail later, but equation 1.3 shows, in a simple way, how each technique relates to the general optical response of a material.

1.3 Techniques Involving $\chi^{(1)}$

Laser light scattering (LLS) depends on $\chi^{(1)}$ via the reflectivity, as does photoluminescence (PL) via the absorption of light. The other techniques which involve $\chi^{(1)}$ can be classified by whether the sample interface has to be changed by formation or removal of a (heterogeneous) adsorbed layer. Spectroscopic ellipsometry (SE), which measures the ratio of (complex) reflection coefficients for s- and p-polarised light (see below), does not require such a change in interface conditions. Reflection anisotropy spectroscopy (RAS) measures the difference in reflectance of light linearly polarised along the two principal axes in the interface plane at near-normal incidence. The interface region must be optically anisotropic and the bulk isotropic, and only azimuthal rotation of the sample or rotation of the optical plane of polarisation is required [6-10]. Both surface differential reflectance (SDR) [11-13] and surface photoabsorption (SPA) [14-19] measure the difference in reflectance when the (surface and) interface layer is changed. SPA uses p-polarised light at or near the Brewster angle (where bulk reflectivity for p-polarised light is a minimum) to increase interface sensitivity.

SE RAS, SDR and SPA are complementary techniques which probe the surface linear susceptibility in different ways, and it is useful to examine their relationship in more detail by considering the linear optical response of a *three-phase system* (or *three-layer system*) comprising vacuum (medium 0), thin film (medium 1) and substrate (medium 2) [20]. If media 1 and 2 are both optically anisotropic with misaligned axes, the optical response is difficult to evaluate. Lekner has considered various cases of crystals and layers recently, and published analytic expressions [21-26], but an isotropic bulk response (medium 2) will be

Table 1.1. Epioptic techniques. The terminology used is shown in **bold**.

Technique	χ-dependence	Abbreviation	
Laser light scattering	$\chi^{(1)}$	LLS	} same
Diffuse elastic light scattering	$\chi^{(1)}$	DELS	
Spectroscopic ellipsometry	$\chi^{(1)}$	SE	
Reflection difference	$\chi^{(1)}$	RD	
Surface differential reflectance	$\chi^{(1)}$	SDR	} same
Surface reflectance spectroscopy	$\chi^{(1)}$	SRS	
Surface photoabsorption	$\chi^{(1)}$	SPA	
Reflection difference spectroscopy	$\chi^{(1)}$	RDS	
Reflection anisotropy spectroscopy	$\chi^{(1)}$	RAS	} same
Surface-induced optical anisotropy	$\chi^{(1)}$	SIOA	
Photoluminescence	$\chi^{(1)}$	PL	
Raman spectroscopy		RS	
- deformation potential	$\partial\chi^{(1)}/\partial Q$		
- Fröhlich interaction	$\partial^2\chi^{(1)}/\partial Q\partial E(0)$		
Optical second harmonic generation	$\chi^{(2)}$	SHG	
Optical sum frequency generation	$\chi^{(2)}$	SFG	

assumed here. For reflection at an interface between medium 1 and medium 2, both the tangential components of the electric and magnetic fields, and the normal components of the electric displacement and magnetic flux density, are required to be continuous across the interface. The normal component of the EM field is thus assumed to be fully screened at the interface. The geometry of interface is shown in Fig. 1.2, where p-polarisation describes an electric vector in the plane of incidence (the xz-plane), s-polarisation an electric vector in the y-direction. The angle of incidence is ϕ_0. Applying the boundary conditions at the interface yields the well-known Fresnel equations for the (complex) amplitude of the reflected wave, which is expressed in normalised form as the complex *reflection coefficient* $r_{12} = E_{12}^r / E_{12}^i$, the ratio of reflected to incident electric field *amplitude* [27]. The *reflectance*, $R_{12}(\omega) = |r_{12}|^2$, is then the ratio of the reflected to incident beam *intensity*. Care is required here because the use of these two terms is sometimes reversed in the literature.

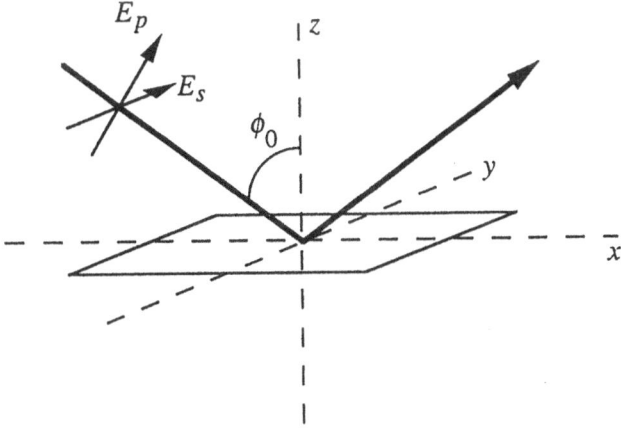

Fig. 1.2. Interface geometry

If the interface is between two biaxial crystals with dielectric tensor components ε_{ixx}, ε_{iyy}, and ε_{izz}, and $i = 1, 2$, with the crystal axes aligned with the laboratory axes, then the Fresnel equations are:

$$r_{12} = \frac{Z_1 - Z_2}{Z_1 + Z_2} \qquad (1.4)$$

where, for s-polarisation,

$$Z_i = \sqrt{\varepsilon_{iyy}} \, \cos \phi_i$$
$$\sqrt{\varepsilon_{1yy}} \, \sin \phi_1 = \sqrt{\varepsilon_{2yy}} \, \sin \phi_2 \qquad (1.5)$$

and ϕ_i is the angle in medium i , and for p-polarisation,

$$Z_i = \cos \phi_i / \sqrt{\varepsilon_{ixx}}$$

$$\sqrt{\varepsilon_{1zz}} \sin \phi_1 = \sqrt{\varepsilon_{2zz}} \sin \phi_2 \qquad (1.6)$$

With this formalism established, a three-phase system can now be investigated, where medium 0 is vacuum, and medium 1 is of thickness d. The reflection coefficient becomes [28, 29]:

$$r_{012} = \frac{r_{01} + r_{12}e^{-i2\beta}}{1 + r_{01}r_{12}e^{-i2\beta}} \qquad (1.7)$$

where the time-dependence of the field is taken as $e^{+i\omega t}$, and β, the phase change on a single pass through medium 1, is given by

$$\beta = (2\pi d / \lambda) \cos \phi_1 \sqrt{\varepsilon_{1jj}} \qquad (1.8)$$

where $j = x, y$, depending on the polarisation. In applying epioptics to surfaces, interfaces and thin films, the condition $d \ll \lambda / 4\pi \cos \phi_1 \sqrt{\varepsilon_{1jj}}$ is always satisfied, which allows the exponential term in equation 1.7 to be replaced by the first order approximation, $(1 - i2\beta)$ [28]. In addition, medium 2 will be a bulk semiconductor for the systems of interest in this book, allowing the replacement of ε_{2jj} by the single component, ε_b, the isotropic bulk value. This leads to the following expressions [29] for the complex reflection coefficient of a thin, biaxial layer of thickness d, on an isotropic substrate in vacuum, where the z-axis of the film is assumed to be perpendicular to the interface, and the x-axis of the film is at an azimuthal angle, ψ, with respect to the plane of incidence:

$$\frac{r_{ss}(\phi_0, \psi)}{r_{ss}^0} = 1 + \frac{i4\pi d \cos \phi_0}{\lambda} \frac{(\varepsilon_b - \bar{\varepsilon} - \Delta\varepsilon \cos 2\psi)}{\varepsilon_b - 1}$$

$$r_{ps}(\phi_0, \psi) = -r_{sp}(\phi_0, \psi)$$

$$= \frac{i4\pi d \cos \phi_0}{\lambda} \frac{\Delta\varepsilon \sin 2\psi \sqrt{\varepsilon_b - \sin^2 \phi_0}}{(\varepsilon_b \cos \phi_0 + \sqrt{\varepsilon_b - \sin^2 \phi_0})(\cos \phi_0 + \sqrt{\varepsilon_b - \sin^2 \phi_0})}$$

$$\frac{r_{pp}(\phi_0, \psi)}{r_{pp}^0} = 1 + \frac{i4\pi d \cos \phi_0}{\lambda(\varepsilon_b - 1)(\varepsilon_b \cos^2 \phi_0 - \sin^2 \phi_0)} [(\varepsilon_b - \bar{\varepsilon})\varepsilon_b$$

$$- (\{\varepsilon_b^2 / \varepsilon_{zz}\} - \bar{\varepsilon}) \sin^2 \phi_0 + \Delta\varepsilon(\varepsilon_b - \sin^2 \phi_0) \cos 2\psi] \qquad (1.9)$$

where the subscripts refer to s- and p-polarised light reflected from (first subscript), and incident on (second subscript), the sample. In these equations, $\bar{\varepsilon} = (\varepsilon_{xx} + \varepsilon_{yy})/2$, $\Delta\varepsilon = (\varepsilon_{yy} - \varepsilon_{xx})/2$, and r_{ss}^0, r_{pp}^0 are the complex reflection coefficients of the bulk:

$$r_{ss}^0 = \frac{\cos\phi_0 - \sqrt{\varepsilon_b - \sin^2\phi_0}}{\cos\phi_0 + \sqrt{\varepsilon_b - \sin^2\phi_0}}$$

$$(1.10)$$

$$r_{pp}^0 = \frac{\varepsilon_b \cos\phi_0 - \sqrt{\varepsilon_b - \sin^2\phi_0}}{\varepsilon_b \cos\phi_0 + \sqrt{\varepsilon_b - \sin^2\phi_0}}$$

These equations assume a step-like change in the dielectric function between phases, and also neglect non-locality. This is discussed further in Sect. 2.2.1. The equations reduce to previously published three-phase model expressions for an isotropic film [29], and for aligned axes [30]. In addition, for the aligned axes case, they are easily rearranged to produce equations in $\Delta\varepsilon_{jj} \equiv (\varepsilon_{jj} - \varepsilon_b)$, and $\Delta\varepsilon_{jj}^{-1} \equiv (\varepsilon_{jj}^{-1} - \varepsilon_b^{-1})$, which are compatible with more general surface models where these terms are treated as integrals, over the selvedge region, of the difference between surface and bulk properties, thus allowing the possibility of including inhomogeneity and local-field effects in the analysis [30-34].

Three general comments can now be made, based on equation 1.9. Firstly, the contribution from the interface region depends on the *difference* between the interface and bulk dielectric response, or on the interface *anisotropy*. Secondly, Hingerl *et al* [29] pointed out recently that the contribution of ε_{zz} to r_{pp} is of the order $|1/\varepsilon_{zz}|$ relative to the leading terms in equation 1.9, because the continuity condition on the displacement field used in deriving equation 1.9 requires that the normal component of the EM field be fully screened, as discussed above. If $|\varepsilon_{zz}|$ is comparable in size to $|\varepsilon_b|$, then the main contribution to r_{pp} comes from ε_{xx}. *The linear optical response from interfaces is expected to be dominated by ε_{xx} and ε_{yy}.* Thirdly, equation 1.9 shows that, at optical wavelengths, the contribution from single atomic layers should be detectable, as reflectivity changes below 10^{-4} are measurable.

Returning now to the various techniques, RAS measures the difference between the (near) normal incidence values of r_{pp} for $\psi = 0°$ and $\psi = 90°$ [6,7]:

$$\left.\frac{\Delta r'}{r'}\right|_{RAS} = 2\frac{r_{pp}(0°,90°) - r_{pp}(0°,0°)}{r_{pp}(0°,90°) + r_{pp}(0°,0°)} \equiv \frac{\Delta r}{r} + i\Delta\theta = \frac{i4\pi d}{\lambda}\frac{(\varepsilon_{xx} - \varepsilon_{yy})}{\varepsilon_b - 1} \quad (1.11)$$

In SDR [11-13], the difference in reflectance is measured when the interface is changed from state a to state b, with a principal axis of the sample aligned along the plane of incidence,

$$\left.\frac{\Delta R_{pp}}{R_{pp}}\right|_{SDR}^{\phi_0} \equiv \frac{R_{pp}^b - R_{pp}^a}{R_{pp}^a} \tag{1.12}$$

where $R_{pp}^i = |r_{pp}^i(\phi_0, \psi)|^2$, with $i = a, b$ and $\psi = 0°$ or $90°$. SPA is a special case of SDR with ϕ_0 chosen to be at or near the Brewster angle of the substrate, as mentioned above [14-19].

Equations 1.11 and 1.12 are closely related. Direct comparison becomes possible if the RAS experiment is performed on interface a and interface b, with both $\Delta r/r$ and $\Delta\theta$ being measured, and if the SDR experiment is performed for each principal axis. Comparison is easiest for normal incidence SDR, if the quantity

$$\left.\frac{\Delta R}{R}\right|_{SDR}^{0°} = \frac{R^b - R^a}{R^b + R^a} \tag{1.13}$$

is measured. For SPA, a conversion factor involving ϕ_0 and ε_b is required to allow comparison of RAS and SPA data [28]. A full SDR or SPA study provides more information than a conventional RAS study because, in principal, SDR may detect optical transitions which are either isotropic and anisotropic in the surface plane, whereas RAS only detects the latter. These techniques are discussed in detail in Chap. 4.

Finally, we return to SE, which is the most complicated of the techniques and appears in a variety of experimental configurations (see Chap. 3). The change in polarisation on reflection is measured as the (complex) ratio of the p- to s-polarised reflection coefficients, and is expressed, historically, in terms of the ellipsometric angles, Δ and Ψ:

$$\rho = r_p/r_s = \tan \Psi \exp i\Delta \tag{1.14}$$

The expression for the bulk response, $\rho^0 = r_{pp}^0/r_{ss}^0$, is simple, but the interface cross-polarisation terms, $r_{ps} = -r_{sp}$, result in the measured value of ρ becoming dependent on the detail of the experimental method. For the most widely used rotating-analyser technique, equation 2.14 becomes [29]:

$$\rho = (r_{pp} + r_{ps} \tan P)/(r_{ss} + r_{sp} \cot P) \tag{1.15}$$

where the linearly polarised incident wave has an input field vector at the polariser angle, P, to the plane of incidence. Equation 1.9 then yields:

$$
\begin{aligned}
\frac{\rho}{\rho^0} = 1 + &\left\{\frac{i4\pi d\cos\phi_0}{\lambda(\varepsilon_b - 1)(\varepsilon_b \cos^2\phi_0 - \sin^2\phi_0)}\right\}\Big[(\varepsilon_b - \bar{\varepsilon})(\varepsilon_b + 1)\sin^2\phi_0 \\
&- \left(\left\{\varepsilon_b^2/\varepsilon_{zz}\right\} - \bar{\varepsilon}\right)\sin^2\phi_0 + 2\Delta\varepsilon\Big(\csc 2P(\varepsilon_b - \sin^2\phi_0) \\
&- \cot 2P\sin^2\phi_0(\varepsilon_b - \sin^2\phi_0)^{1/2}\Big)\sin 2\psi \\
&+ \Delta\varepsilon(\varepsilon_b\{1 + \cos^2\phi_0\} - 2\sin^2\phi_0)\cos 2\psi\Big]
\end{aligned}
\tag{1.16}
$$

This (complicated) expression has the same general features as equation 1.9, showing that SE, combined with sample rotation, can provide the same information as RAS and SPA.

1.4 Raman Scattering

The second and third terms in equation 1.3 describe Raman spectroscopy (RS), where optical phonons are created and annihilated producing a frequency shift. The underlying theory of the Raman scattering process is well understood both in classical and quantum terms [35]. The main epioptic technique is phonon RS, although other single particle and collective excitations with smaller cross-sections can be important in two-dimensional electron gas systems [5]. In outline, the incident radiation induces a dipole which will vary with some deformation coordinate, Q, representing the displacement of the ion cores of the system. This deformation potential may produce elastic (Rayleigh) scattering, or inelastic (Raman) scattering where a phonon is created (Stokes process) or destroyed (Anti-Stokes process). The Raman process is highly non-specular and has a small cross-section (typically one Raman photon per 10^{10} incident photons). It is only recently that instrumental advances have allowed RS to demonstrate its potential as a surface and interface probe. The Raman terms in Table 1.1 indicate that the Raman effect is governed by third and fourth rank tensors. Crystal and molecular symmetry reduces the number of non-zero tensor components, giving rise to selection rules, while the magnitude of non-zero components can be determined from a perturbative approach. The Raman effect can be greatly enhanced by both local electric fields and electronic resonances at the excitation frequency, but this may involve a reduction in the amount of symmetry information available. RS is thus one of two main epioptic techniques which provide vibrational information, the other being sum frequency generation (SFG).

1.5   Second Harmonic and Sum Frequency Generation

The fourth term in equation 1.3, which has a polarisation quadratic in the EM field, describes the lowest order nonlinear optical response which produces second harmonic generation (SHG). More generally, the $\chi^{(2)}$ term describes three wave mixing, including sum frequency generation (SFG), and is written $\chi^{(2)}(-\omega_3; \omega_1, \omega_2)$, where ω_1 and ω_2 are the incoming field frequencies, ω_3 is the outgoing field frequency, and $\omega_3 = \omega_1 \pm \omega_2$ [4]. Considering other nonlinearities briefly, third harmonic generation, in the absence of any strong resonantly-enhanced surface electric dipole effects, is a bulk probe, while higher-order nonlinearities are too small to be useful. SHG and other three-wave mixing phenomena are potentially surface sensitive at non-destructive power densities [36]. This is most easily seen for centrosymmetric materials, where the electric dipole term is parity forbidden, leaving only higher order contributions (magnetic dipole and electric quadrupole effects). At a surface or interface, the bulk symmetry is broken and electric dipole effects are allowed.

The surface-dipole-allowed contribution to SHG, arising from the nonlinear polarisation field, $P_i(2\omega)$, induced by an incident laser field at frequency ω, can be written as (equation 1.1, second term):

$$P_i(2\omega) = \varepsilon_0 \left\{ \chi^s_{ijk} E_j(\omega) E_k(\omega) \right\} \tag{1.17}$$

where χ^s_{ijk} is the second-order susceptibility tensor component reflecting the structure and symmetry properties of the interface. It is now easily seen that the bulk electric dipole contribution from a centrosymmetric material is parity-forbidden: under a parity operation the fields change sign, but a centrosymmetric χ^s_{ijk} component does not, leading to only the left side of equation 1.17 changing sign.

The tensor can have, in principle, 18 independent components, but any symmetry elements present in the surface will reduce this number substantially. Choice of experimental geometry and polarisation vectors based on the point group symmetry of the surface can then be used to isolate *individual* tensor components, and structural information can readily be obtained. Symmetry analysis is a powerful tool here and it should be noted that the symmetry rules for the linear and nonlinear response are different. Cross-sections for three-wave mixing events are small, with typically one signal photon per 10^{13}-10^{17} incident photons and, with large field gradients normal to the surface, higher order SH signals can be generated in the bulk of a semiconductor crystal which are comparable in size to the interface signal. However, where an interface resonance can be probed, the dipolar contribution from the interface may dominate the SH response. This spectroscopic aspect of SFG [37] and SHG [38] is now being exploited.

1.6 Application to Semiconductor Growth

This is a rapidly developing area, with major advances being made using linear optical techniques. For example, *in situ* monitoring of layer-by-layer growth in vapour phase epitaxy (VPE) reactors is now possible using epioptic techniques.

Table 1.2. Epioptics applied to thin film growth on (001) semiconductor surfaces

Application	Group-IV (001) surfaces: Si, Ge and SiGe								
	LLS	SE	SDR	RAS	SPA	PL	RS	SHG	SFG
Substrate characterisation (as-received)		•				••			
Oxide interface characterisation		••						••	••
Oxide desorption		••	•	•	•		•	••	••
Surface crystallographic structure							•	••	••
Surface electronic structure		•	••		••	••		•	••
Surface vibrational structure		•				•	••		••
Surface stoichiometry (Si_xGe_{1-x})		••	•	••	•		•	•	•
Surface morphology and roughness	••	••					•	•	•
Surface strain		••					••	•	•
Growth rate	•	•		••	•			•	•
Growth mode	•	••		•			•	••	••
Growth temperature		••					••	•	•

III-V and II-VI (001) surfaces									
	LLS	SE	SDR	RAS	SPA	PL	RS	SHG	SFG
Substrate characterisation (as received)		•		••		••			
Oxide interface characterisation		••		•	•			•	••
Oxide desorption		••	•	••	•		••	••	••
Surface crystallographic structure				••			••	••	••
Surface electronic structure		•	••	•	••	••	•	•	••
Surface vibrational structure		•				•	••		••
Surface stoichiometry		••	••	••	••		••	•	•
Surface morphology	••	••	•	••	••		••	•	•
Surface strain		••					••	•	•
Growth rate	•	••		••	••		••	•	•
Growth mode	•	••		••	••		••	•	•
Growth temperature		••		•	•	•	••	•	•

• some information available   •• good information available

In VPE reactors, the gas pressure prevents the use of conventional surface science techniques, such as reflection high energy electron diffraction (RHEED) which is widely exploited in molecular beam epitaxy (MBE). The application of epioptics to semiconductor growth will be discussed later, in more detail, for each technique, but here we list, in Table 1.2, our view of the potential of epioptics in the various areas of thin film semiconductor growth and characterisation. Epioptic techniques have wide applicability for III-V and II-VI semiconductors, and are somewhat more restricted for elemental semiconductors. Although Table 1.2 refers to the most common (001) surfaces, most of the summary would apply to (111) surfaces.

References

1.   McGilp, J.F.: Journal of Physics: Condensed Matter 2, 7985 (1990)
2.   McGilp, J.F.: Journal of Physics: Condensed Matter 1, SB85 (1989)
3.   Debe, M.K.: Progress in Surface Science *24*, 1 (1987)
4.   Hopf, F.A., Stegeman, G.I.: Applied classical electrodynamics, vols. 1 and 2. Wiley, New York 1986
5.   Geurts, J., Richter, W.: Springer Proceedings in Physics, vol. 22. Springer, Berlin 1987, p. 328
6.   Aspnes, D.E.: Journal of Vacuum Science and Technology *B 3*, 1498 (1985)
7.   Berkovits, V.L., Makarenko, I.V., Minashvili, T.A., Safarov, V.I.: Solid State Communications *56*, 449 (1985)
8.   Aspnes, D.E., Harbison, J.P., Studna, A.A., Florez, L.T.: Physical Review Letters *59*, 1687 (1987)
9.   Paulsson, G., Deppert, K., Jeppesen, S., Jönsson, J., Samuelson, L., Schmidt, P.: Journal of Crystal Growth *111*, 115 (1991)
10. Kamiya, I., Aspnes, D.E., Tanaka, H., Florez, L.T., Harbison, J.P., Bhat, R.: Physical Review Letters *68*, 627 (1992)
11. Chiaradia , P., Chiarotti, G., Ciccacci, F., Memeo, R., Nannarone, S., Sassaroli, P., Selci, S.: Surface Science *99*, 7011 (1980)
12. Chiaradia, P., Cricenti, A., Selci, S., Chiarotti, G.: Physical Review Letters *52*, 1145 (1984)
13. Selci, S., Chiaradia, P., Ciccacci, F., Cricenti, A., Sparvieri, N., Chiarotti, G.: Physical Review *B 31*, 4096 (1985)
14. Kobayashi, N., Horikoshi, Y.: Japanese Journal of Applied Physics *28*, L1880 (1989)
15. Kobayashi, N., Horikoshi, Y.: Japanese Journal of Applied Physics *29*, L702 (1990)
16. Kobayashi, N., Horikoshi, Y.: Japanese Journal of Applied Physics *30*, L319 (1991)
17. Kobayashi, N., Yamauchi, Y., Horikoshi, Y.: Journal of Crystal Growth *115*, 353 (1991)
18. Farrell, T., Armstrong, J.V., Kightley, P.: Applied Physics Letters *59*, 1203 (1991)
19. Nishi, K., Usui, A., Sakaki, H.: Applied Physics Letters *61*, 31 (1992)
20. Azzam, R.M.A., Bashara, N.M.: Ellipsometry and polarized light. North Holland, Amsterdam 1977
21. Lekner, J.: Journal of Physics: Condensed Matter 3, 6121 (1991)
22. Lekner, J.: Journal of Physics: Condensed Matter *4*, 1387 (1992)
23. Lekner, J.: Journal of the Optical Society of America *A 10*, 1579 (1993)
24. Lekner, J.: Journal of the Optical Society of America *A 10*, 2059 (1993)
25. Lekner, J.: Pure and Applied Optics *3*, 307 (1994)
26. Lekner, J.: Pure and Applied Optics *3*, 821 (1994)
27. Ishimaru, A.: Electromagnetic Wave Propagation, Radiation and Scattering. Prentice Hall, Englewood Cliffs 1991

28. McIntyre, J.D.E., Aspnes, D.E.: Surface Science *24*, 417 (1971)
29. Hingerl, K., Aspnes, D.E., Kamiya, I., Florez, L.T.: Applied Physics Letters *63,* 885 (1993)
30. Bagchi, A., Barrera, R.G., Rajagopal, A.K.: Physical Review *B 20*, 4824 (1979)
31. Del Sole, R.: Solid State Communications *37*, 537 (1981)
32. Plieth, W.J., Naegele, K.: Surface Science *64*, 484 (1977)
33. Plieth, W.J., Bruckner, H., Hensel, H.-J.: Surface Science *101*, 261 (1980)
34. Kelly, M.K., Zollner, S., Cardona, M.: Surface Science *285*, 282 (1993)
35. Long, D.A.: Raman Spectroscopy. McGraw-Hill, New York 1977
36. Shen, Y.R.: Nature *337*, 519 (1989)
37. Guyot-Sionnest, P., Hunt, J.H., Shen, Y.R.: Physical Review Letters *59*, 1597 (1987)
38. McGilp, J.F., Cavanagh, M., Power, J.R., O'Mahony, J.D.: Applied Physics *A 59*, 401 (1994)

Chapter 2. The Linear Optical Response

Rodolfo Del Sole,[1] Anatolii Shkrebtii,[1] Jiang Guo-Ping,[1] and Charles Patterson[2]

[1] Dipartimento di Fisica, II Università di Roma, 'Tor Vergata', Via della Ricerca Scientifica, I-00133 Roma, Italia

[2] Department of Physics, Trinity College, Dublin 2, Ireland

2.1 Introduction

The linear optical response theory considered in this chapter relates to the experimental techniques of reflectance anisotropy spectroscopy (RAS) [1-6] and surface differential reflectance (SDR) [7-15]. These experimental techniques were outlined in Ch. 1, and are described in detail in Ch. 4. This chapter describes calculations of the RAS and SDR spectra from clean and adsorbate-covered Si, GaAs and GaP surfaces. The experiments aim to extract information about how the narrow selvedge region at a surface contributes to the reflected electromagnetic field in vacuum. Thus, the field amplitudes in this region at the surface are a very important ingredient in the calculation. What provides an interesting challenge for the theorist in this problem is that the region where the field amplitudes need to be known accurately is also the region where they are changing very rapidly (over ~1 nm) from their vacuum values to the bulk values. We recognise three ill-defined regions (in the sense that their edges are not sharp), which are important in determining the optical response of a surface: I the vacuum region; II the (narrow) selvedge region; III the bulk. Since the advent of modern surface science, probably the first attempt to include these three regions was in the work of McIntyre and Aspnes [16]. This has come to be known as the three-phase model (see Sect. 1.3).

Finding the field amplitudes at the surface of a semiconductor is made considerably more difficult by the fact that the EM field to which an electron is responding is modified by the polarisation of nearby electrons. The total microscopic EM field is the sum of the applied field and the field induced by the polarisation of other electrons. This field is referred to as the *local field* and it varies rapidly on the scale of an interatomic distance inside a solid.

Perhaps the simplest model which can be used to convey the idea of a local field effect in real space is that of a solid consisting of 'atoms' which are infinitesimal or point-like polarisable entities which create point dipole electric fields when a dipole is formed at the atom. In such a solid, the atoms are coupled by point dipole electric fields, and the local field at an atom is the applied field plus the point dipole fields, set up by induced dipoles at all other sites. Such a model may seem very naive at first sight, but it has been used successfully to account for excitonic behaviour in molecular crystals [17, 18], and reflectance at semiconductor surfaces [19, 20]. In these cases the model has been successful, probably because the polarisable units chosen were sufficiently far apart that the point dipole fields are a reasonable approximation to the actual fields. In the case of excitonic behaviour in molecular crystals [17, 18], the polarisable entity was a large organic molecule. In the work that is described in a later section of this chapter, cluster calculations are used to calculate interactions between neighbouring bonds in Si from first principles, so that the method of coupled, polarisable bonds can be used to study semiconductor surfaces.

2.2 Optical Formalism

Three steps are involved in the theory: (i) the determination of one-electron wave functions; (ii) the calculation of the dielectric susceptibility, possibly including many-body interactions; (iii) the solution of light-propagation equations. After the pioneering work of Feibelman [21], the last problem was solved in a quite simple way by Bagchi, Barrera and Rajagopal [22]. The surface contribution to reflectivity can be calculated, according to the formulas given below, from the dielectric susceptibility. In turn, this can be calculated, within the one-electron or random phase approximation (RPA) from single-particle wave functions [23]. These might be calculated from first principles using the Green's function method and the GW approximation to account for electron exchange and correlation [24]. Unfortunately, such calculations are presently too demanding, from the computational point of view, to be applied to the optical properties of most surfaces. Therefore, one has to resort to more empirical methods. Two of them, namely the tight-binding and the self-consistent pseudopotential method, have been implemented for the optical properties of surfaces and will be described in the next section.

2.2.1 Reflectance and Ellipsometry

We begin by considering reflectance of a crystal occupying the half space $z > 0$, with a surface region of thickness, d, which has different optical properties compared to the bulk. Bagchi, Barrera and Rajagopal [22] have solved the light propagation equations for jellium to first order in $\omega d/c$ ($= 2\pi d/\lambda$, see Sect. 1.3), where c is the velocity of light. Later, Del Sole [25] solved these equations for a crystalline solid yielding the surface contribution to the reflectance, $\Delta R/R$, where R is the reflectance of a sharp boundary according to the Fresnel equations (see Sect. 1.3) and $\Delta R/R$ is the relative deviation from these formulae. For s-polarised light this is:

$$\Delta R_s / R_s = 4(\omega/c)\cos\phi_0 \, \mathrm{Im}\{\Delta\varepsilon_{yy}/(\varepsilon_b - 1)\} \tag{2.1}$$

for light incident from vacuum in the xz-plane and, for p-polarised light,

$$\Delta R_p / R_p = 4(\omega/c)\cos\phi_0 \, \mathrm{Im}\left\{ \frac{(\varepsilon_b - \sin^2\phi_0)\Delta\varepsilon_{xx} + (\varepsilon_b^2 \sin^2\phi_0)\Delta\varepsilon_{zz}^{-1}}{(\varepsilon_b - 1)(\varepsilon_b \cos^2\phi_0 - \sin^2\phi_0)} \right\} \tag{2.2}$$

Here ϕ_0 is the angle of incidence, ε_b is the bulk dielectric function, and

$$\Delta\varepsilon_{ij} = \int_{-\infty}^{\infty} dz \int_{-\infty}^{\infty} dz' \left[\varepsilon_{ij}(z,z') - \delta_{ij}\delta(z-z')\varepsilon_0(z) \right]$$
$$- \int_{-\infty}^{\infty} dz \int_{-\infty}^{\infty} dz' \int_{-\infty}^{\infty} dz'' \int_{-\infty}^{\infty} dz''' \varepsilon_{iz}(z,z')\varepsilon_{zz}^{-1}(z',z'')\varepsilon_{zj}(z'',z''') \tag{2.3}$$

for $i, j = x, y$ and

$$\Delta\varepsilon_{zz}^{-1} = \int_{-\infty}^{\infty} dz \int_{-\infty}^{\infty} dz' \left[\varepsilon_{zz}^{-1}(z,z') - \left(\delta(z-z')/\varepsilon_0(z) \right) \right]. \tag{2.4}$$

Here $\varepsilon_0(z) = \theta(-z) + \varepsilon_b\theta(z)$ is the zero-order dielectric function yielding Fresnel equations, $\varepsilon_{ij}(z,z')$ is the non-local macroscopic dielectric tensor [26] of the solid-vacuum interface which, in principle, accounts for all microscopic and many-body effects, and $\varepsilon_{zz}^{-1}(z,z')$ is the inverse kernel of $\varepsilon_{zz}(z,z')$. The frequency dependence of the surface dielectric tensor and the bulk dielectric function is suppressed for simplicity. Similar formulae are obtained for internal reflectivity, namely for the reflectivity of light coming from inside the sample [25]:

$$\Delta R_s^{\mathrm{int}} / R_s = -4(\omega/c)\varepsilon_b^{1/2}\cos\phi_0 \, \mathrm{Im}\{\Delta\varepsilon_{yy}/(\varepsilon_b - 1)\} \tag{2.5}$$

$$\Delta R_p^{int}/R_p = -4(\omega/c)\varepsilon_b^{1/2}\cos\phi_0 \, \text{Im}\left\{\frac{(\varepsilon_b \sin^2\phi_0 - 1)\Delta\varepsilon_{xx} - (\varepsilon_b \sin^2\phi_0)\Delta\varepsilon_{zz}^{-1}}{(\varepsilon_b - 1)(\varepsilon_b \sin^2\phi_0 - \cos^2\phi_0)}\right\}.$$

$$(2.6)$$

For a microscopic evaluation of surface optical properties one has to determine equations 2.3 and 2.4 from wave functions. Since many-body effects are difficult to incorporate, they are neglected. In this case, $\varepsilon_{ij}(z,z')$ is the RPA dielectric tensor, which can be computed from the one-electron wave functions. It is usually evaluated mimicking the semi-infinite crystal by a slab of several (from 10 to 30) atomic layers. The second term in equation 2.3 is more difficult to evaluate than the first one. It originates from the off-diagonal terms $\varepsilon_{iz}(\omega,z,z')$ of the dielectric susceptibility of the semi-infinite crystal. Such terms vanish in the infinite crystal because of the cubic symmetry. In many cases, they also vanish at surfaces by symmetry: this occurs when there is a symmetry operation of the semi-infinite crystal under which $p^i (i = x, y)$ and p^z transform in different ways. In other cases, they can be approximately evaluated according to the anisotropic three-phase model (Sect. 1.3). They are small in all cases considered so far [27] and will be neglected from now on. The first term in equation 2.3 is related to the half-slab (hs) polarisability:

$$\alpha_{ij}^{hs}(\omega) = (1/8\pi)\int_{-\infty}^{\infty}dz\int_{-\infty}^{\infty}dz'\left[\varepsilon_{ij}^{slab}(z,z') - \delta_{ij}\delta(z-z')\right] \qquad (2.7)$$

which has dimension of length because of the twofold integral over z. The s-wave surface optical properties can be expressed as [27]:

$$\Delta R_s/R_s = 4(\omega/c)\cos\phi_0 \, \text{Im}\{4\pi\alpha_{yy}^{hs}/(\varepsilon_b - 1)\} \qquad (2.8)$$

in terms of α^{hs}. Within the single-particle scheme, the imaginary part of $\alpha_{ij}^{hs}(\omega)$ is simply related to the transition probability induced by the radiation between slab states:

$$\text{Im}\{4\pi\alpha_{ii}^{hs}(\omega)\} = (4\pi^2 e^2/m^2\omega^2 A)\sum_{\mathbf{k}}\sum_{v,c}\left|p_{v,c}^i(\mathbf{k})\right|^2 \delta(E_c(\mathbf{k}) - E_v(\mathbf{k}) - \hbar\omega) \quad (2.9)$$

where $p_{v,c}^i(\mathbf{k})$ is the matrix element of the i-component of the momentum operator between initial (v) and final (c) slab states at the point, $\mathbf{k}$, in the two-dimensional Brillouin Zone (BZ), and A is the sample area. The real part is computed via Kramers-Krönig transforms. One more ingredient of equations 2.1 and 2.2 is the bulk dielectric function, whose imaginary part is analogous to equation 2.9, where eigenstates and eigenvalues of the infinite crystal, together with three-dimensional k-vectors, are involved. The study of the optical response of surfaces within the single particle picture amounts, then, to calculating i) the single particle spectrum, and ii) the transition probability between occupied and unoccupied states, both for the perfect crystal and for the crystal with a surface.

In ellipsometry, the response is generally described in terms of the reflection coefficients of s- and p-polarised light, as discussed in Sects. 1.3 and 3.1. The surface contribution to the ellipsometric parameter, ρ (defined in Sect. 1.3), in the cases where the surface dielectric tensor is diagonal, is given by [28]:

$$\frac{\Delta\rho}{\rho} = -2i(\omega/c)\cos\phi_0 \times$$

$$\frac{(\varepsilon_b - \sin^2\phi_0)\Delta\varepsilon_{xx} + (\sin^2\phi_0 - \varepsilon_b\cos^2\phi_0)\Delta\varepsilon_{yy} + (\varepsilon_b^2\sin^2\phi_0)\Delta\varepsilon_{zz}^{-1}}{(\varepsilon_b - 1)(\varepsilon_b\cos^2\phi_0 - \sin^2\phi_0)}$$

$$(2.10)$$

In the absence of surface effects the Fresnel equations, which give ρ^0 (see Sect. 1.3) in terms of ε_b, can be inverted, yielding:

$$\varepsilon_b = \sin^2\phi_0\{1 + \tan^2\phi_0[(1-\rho^0)/(1+\rho^0)]^2\} \qquad (2.11)$$

Usually, ellipsometric measurements are interpreted within the two-layer model, i.e. neglecting the surface layer and assuming that an effective homogeneous medium occupies the half space $z > 0$, as discussed in detail in Sect. 3.3. The effective dielectric function, $<\varepsilon>$, is obtained from equation 2.11 by using ρ instead of ρ^0. With the help of equations 2.10 and 2.11 we find:

$$<\varepsilon> = \varepsilon_b - 4\sin^2\phi_0\tan^2\phi_0[\rho^0(1-\rho^0)/(1+\rho^0)^3]\Delta\rho/\rho \qquad (2.12)$$

The second term on the right-hand side, which accounts for surface effects, is of the order of d/λ: therefore $<\varepsilon>$ and ε_b differ by a few percent.

2.2.2 The Anisotropic Three-Phase Model

In 1971, McIntyre and Aspnes began using the isotropic three-phase model [16] to describe the optical properties of surface layers, as discussed in Sect. 1.3. This model neglected the *non-locality* and *anisotropy* of the surface dielectric tensor, and approximated its inhomogeneity by a two-step function. This model remains very popular because of its simplicity, and has recently been extended to include surface anisotropy (Sect. 1.3).

The general solution of light-propagation equations, leading to equations 2.1, 2.2, 2.5, 2.6 and 2.10, allows us to go beyond this model and compute the s-polarised reflectance directly from wave functions. However, it is still very difficult to calculate the p-polarised reflectance, and therefore ellipsometric results, because of the inversion $\varepsilon_{zz}(z,z')$ required to evaluate $\Delta\varepsilon_{zz}^{-1}$. This can be achieved within the anisotropic three-phase model, where the surface dielectric function, although *local* and diagonal, has ε_{xx}, ε_{yy} and ε_{zz} components. Within this model:

$$\Delta\varepsilon_{ii} = d(\varepsilon_{ii} - \varepsilon_b) \qquad \text{for } i = x, y \qquad (2.13)$$

and
$$\Delta\varepsilon_{zz}^{-1} = d(\varepsilon_{zz}^{-1} - \varepsilon_b^{-1}). \qquad (2.14)$$

These quantities are used in equations 2.1, 2.2, 2.5, 2.6 and 2.10 to yield the reflectance, and the ellipsometric parameter, ρ.

To use the three-phase model in conjunction with microscopic slab calculations, the surface dielectric tensor must be determined in terms of the slab polarisability, $2\alpha_{ii}^{hs}$ ($i = x, y, z$ from here on). If the dielectric tensor is assumed to be local, and the three-phase model is valid, then for each surface of a slab of thickness, L :

$$2\alpha_{ii}^{hs} = 2d\alpha_{ii} + (L - 2d)\alpha_b \qquad (2.15)$$

where α_{ii} and α_b are the surface layer and bulk polarisabilities, respectively. Given a reasonable choice of d, we can solve equation 2.15 for α_{ii}. The choice of d is dictated by the requirement of a very thin surface layer, constrained by the fact that, if d is too small, the imaginary part of α_{ii} would be negative. A choice of d in the range of a few monolayers, say 0.5 nm, usually satisfies both requirements. The surface dielectric tensor, to be used in equations 2.13 and 2.14, is, of course, related to the surface layer polarisability by:

$$\varepsilon_{ii} = 1 + 4\pi\alpha_{ii}. \qquad (2.16)$$

It is remarkable that the *s*-polarised reflectance is correctly given, according to equation 2.8, for any choice of d. In fact, two different d-values will yield, from equation 2.15, two different α_{ii}, whose *difference* is proportional to α_b, i.e. to $(\varepsilon_b - 1)$. Therefore, when they are used in equation 2.8, their difference will yield a real number, which will not contribute to the imaginary part to be used in the equation.

2.3 Slab Calculations

2.3.1 Tight-binding Calculations

We outline a method for the evaluation of surface reflectance which has been applied to a number of Si surfaces. We use bulk bands calculated using the tight-binding Hamiltonian of Vogl *et al* [29] and scale the Hamiltonian matrix, H, elements according to the $1/d^2$ rule to account for the changes in the distance of neighbouring atoms. We calculate the electronic states and the optical properties of slabs of several (from 12 to 31) atomic layers, assuming some model for the surface relaxation or reconstruction, according to equations 2.8 and 2.9 of the previous section. A slightly different method has been recently developed by Chang and Aspnes [3]. Since the wave functions are expanded in local orbitals, we need to calculate the momentum matrix elements, $\mathbf{p}$, between orbitals n and n' in cells $\mathbf{R}$ and $\mathbf{R}'$, respectively. We calculate these matrix elements using the commutator:

$$\mathbf{p} = i(m/\hbar)[\mathrm{H}, \mathbf{r}] \qquad (2.17)$$

and neglecting the matrix elements of the position operator, $\mathbf{r}$, between different atoms. This is possible because the orbitals are orthogonal each other. We have estimated that this approximation leads to errors of less than 10% in the matrix elements of $\mathbf{p}$. The intra-atomic matrix elements are obtained, by a fit of bulk Si optical properties, as $<s|x|p_x> = 0.027$ nm and $<s^*|x|p_x> = 0.108$ nm [30]. It is straightforward to obtain:

$$<n\mathbf{R}|\mathbf{p}|n'\mathbf{R}'> = i(m/\hbar)\sum_{n''} <n\mathbf{R}|\mathrm{H}|n''\mathbf{R}'><n''\mathbf{R}'|\mathbf{r}|n'\mathbf{R}'>$$
$$- <n\mathbf{R}|\mathbf{r}|n''\mathbf{R}><n''\mathbf{R}|\mathrm{H}|n'\mathbf{R}'> \qquad (2.18)$$

2.3.2 Application of the Tight Binding Method to the Si(111)1x1-As Surface

The SDR spectrum is simulated by subtracting the reflectance of a hydrogenated surface from that of the clean surface. In experiments, oxygen chemisorption is usually employed. It is much simpler to carry out calculations for hydrogen- rather than for oxygen-covered surfaces. Approximating the oxidised surface by a hydrogenated one is expected to be valid for frequencies below the bulk gap, where bulk transitions have no role in determining the SDR response. In this case, both hydrogen and oxygen remove the dangling-bond states from the near-gap region. The validity of this approach is still to be ascertained above the gap, where the two terminations could affect bulk states near the surface in different ways.

As an example of the relevance of this technique, we compare the theoretical SDR spectrum [30] for the Si(111)2x1 surface with experiment in Fig. 2.1 [14]. The overall agreement is good. More importantly, the strong asymmetry of the peak near 0.5 eV is reproduced by the calculation, which relies on Pandey's π-bonded chain model [31], while the opposite asymmetry is predicted on the basis of other models [15]: the optical experiment not only has confirmed Pandey's model, but has also ruled out all other reconstruction models formulated so far.

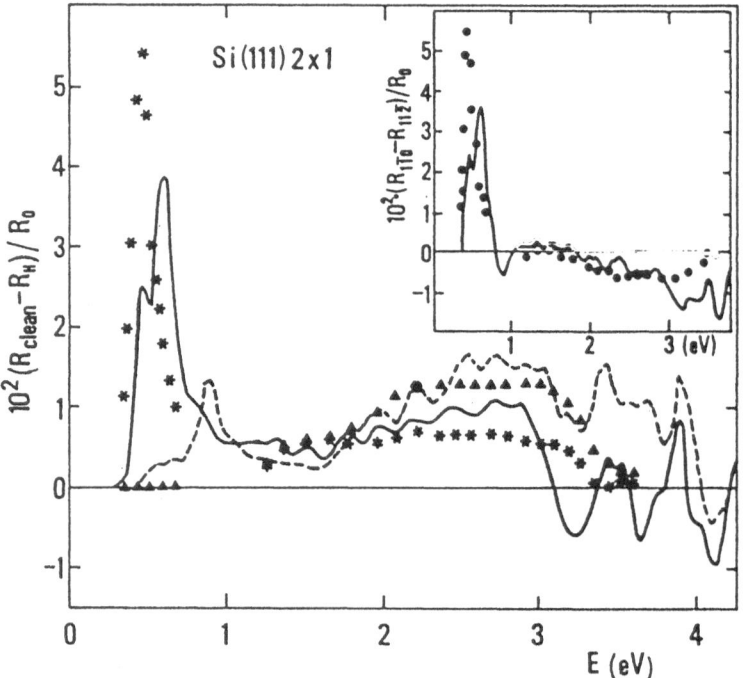

Fig. 2.1. Differential reflectivity of Si(111) 2x1 calculated as the difference between the clean and hydrogen-covered 1x1 surface reflectivities [30]. Full line: y-polarisation (parallel to the chains). Dashed line: x-polarisation (perpendicular to the chains). Stars and triangles are experimental results [14] for y- and x-polarisations, respectively. Inset: RAS $(\Delta R/R)_y - (\Delta R/R)_x$. Full line: calculation; dots: experimental results.

We now describe our work on Si(111) 1x1-As. The atomic and electronic structure of the Si(111) surface covered by one ML of As is presently well understood [32, 33]. Arsenic atoms substitute Si atoms in the first layer, giving rise to doubly occupied As dangling bonds (As has 5 valence electrons instead of 4 in Si). Thus the inherent instability of the ideal Si(111) surface is removed, because the unpaired electrons of that surface are now coupled into a stable lone pair. The first layer relaxes outward by about 0.02 nm [32]. In view of the structural simplicity of this surface, it has been possible to calculate [33] its electronic structure by the most advanced method presently available, namely the GW method, which usually yields energy bands in very good agreement with experiment. For Si(111) 1x1-As, a filled and an empty surface band are found [34]: the filled band is mostly constituted by As lone-pair orbitals, while the empty band is localised mainly on Si-As back antibonds.

Measurement of the optical properties of this surface has presented a challenge to experimentalists, since the techniques used to make SDR and RAS methods surface sensitive cannot be used in its case. The symmetry of the surface prevents

RAS being used, while the As coverage makes the surface stable with respect to oxidation, which is used in SDR experiments. The information about the optical properties of this surface comes from the experiments carried out by Kelly *et al* using SHG [35], and by Rossow *et al* using SE [36]. In order to provide theoretical support for these experiments, we have carried out calculations of the optical properties of this surface. We have computed the electronic structure of a slab of 12 Si (111) layers, where in the first and last layer Si atoms are substituted by As atoms, according to the tight-binding method described in Refs. 29 and 30. The outward relaxation of the surface layers has been properly included.

The resulting surface states are shown in Fig. 2.2. The filled surface state, originating from the doubly occupied As dangling bond, is in good agreement with GW calculations [33] and experiment [34]. We found three empty surface bands along the ΓK direction (K is the corner of the hexagonal Surface Brillouin Zone, SBZ), and two empty bands along the ΓM direction (M is the middle of the edge of the SBZ). Among these, the lowest one, showing $sp_z s^*$ character, can be identified with that of Ref. 33, with which it is in reasonable agreement. The higher bands are, strictly speaking, surface resonances rather than surface states, since they are always overlapping, in energy, with the bulk conduction band. Such bands, which show at least partially the same $sp_z s^*$ character, are also important for optical properties.

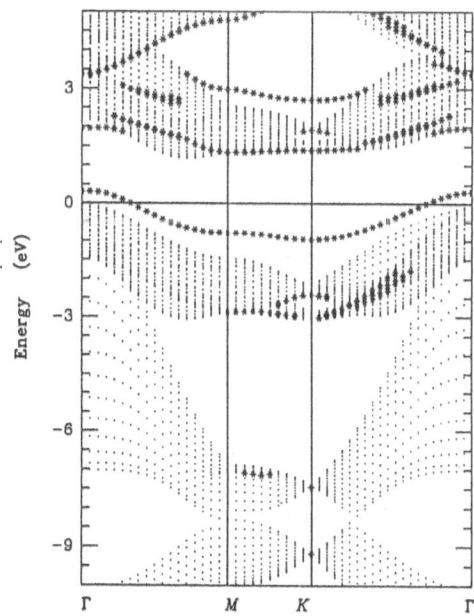

Fig. 2.2. Surface bands of Si(111)1x1-As calculated within the tight-binding approach. Stars indicate surface states or resonances [37].

The joint density of states (JDOS) projected on the first layer is plotted in Fig. 2.3. It is narrower than the total density of states, peaked around 4 eV, with weaker and broader structures around 3 and 5 eV. We expect transitions between

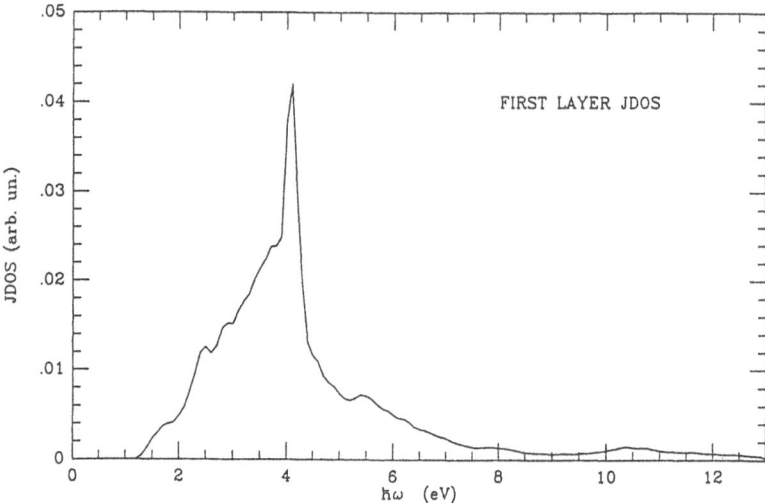

Fig. 2.3. JDOS of Si(111) 1x1-As projected on the first layer, as calculated in the present approach .

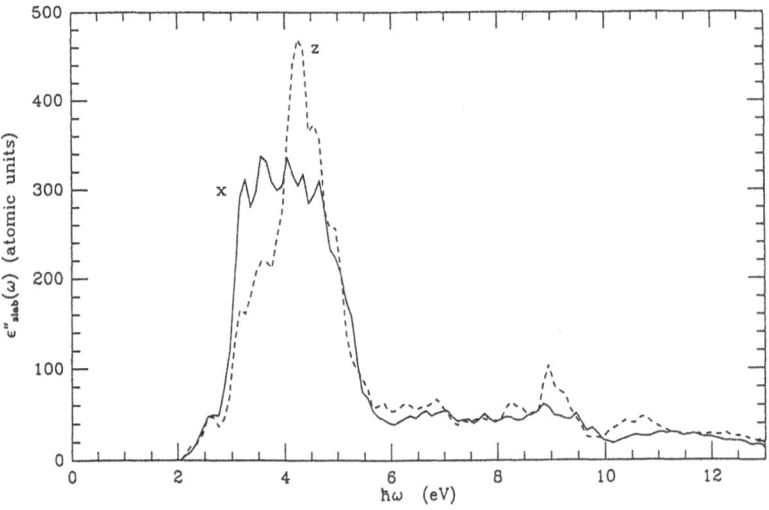

Fig. 2.4. Polarisability of a Si(111) slab of 10 Si and 2 As planes [37].

surface states to occur mainly at these energies. The imaginary part of the slab polarisability in dimension of length is shown in Fig. 2.4 for light polarised along the surface (the xx- and yy-components of the dielectric tensor are equal because of symmetry) and perpendicular to the slab (zz-component of the dielectric tensor). The slab polarisability includes bulk and surface transitions. Since the bulk polarisability is isotropic, the importance of surface transitions can be roughly estimated by the anisotropy of the slab. Such anisotropy is particularly strong between 3 and 5 eV, suggesting that surface-related transitions must be strong in this energy range, in agreement with the previous argument based on the projected joint density of states. The structure above 3 eV is stronger for light polarised in the surface plane. A replica of the peak at 4 eV of the surface joint density of states is seen in the zz component of the slab polarisability. It is related to transitions from the filled As dangling bond to the As-Si back-antibonds. The location, around 4 eV, of strong transitions between surface states points to the importance of the upper empty bands shown in Fig. 2.2. The surface transitions at 3, 4 and 5 eV are more clearly visible in Fig. 2.5, where the imaginary part of the surface layer dielectric function is shown, having been extracted from the slab polarisability according to equation 2.15.

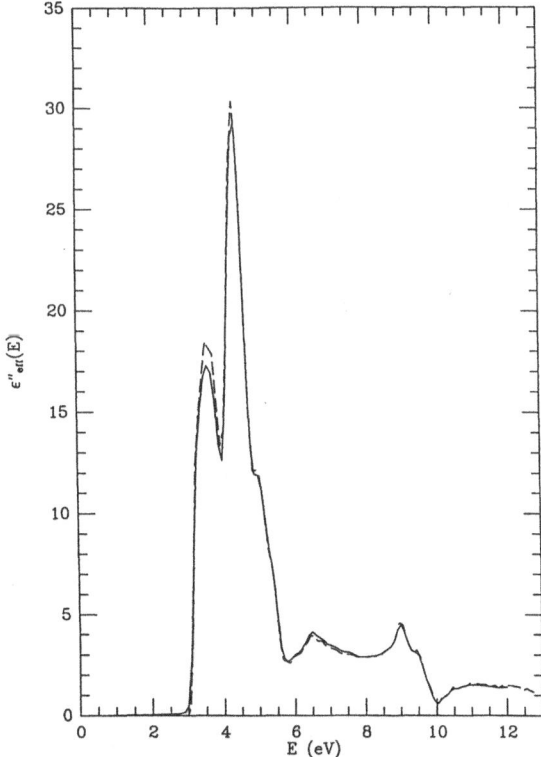

Fig. 2.5. Dielectric tensor of the Si(111) 1x1-As surface within the three-phase model. The surface layer consists of 4 atomic layers [37].

We now consider ellipsometric measurements. We have calculated the effective bulk dielectric function as extracted from ellipsometry using Fresnel equations, i.e. ignoring the specific features of the surface. Since these are present in experiment, the effective dielectric function will embody a surface contribution and will be different from the bulk one, as shown by equation 2.12 of the previous section. The imaginary part of the calculated effective dielectric function is shown by the dashed line in Fig. 2.6. It differs by only a few percent from the bulk one, not shown in the figure; since the latter is determined just within such precision, a direct comparison between them is meaningless. In the SE experiment [36], it has been compared with that of the clean Si(111)7x7 reconstructed surface. This is, however, too difficult for our calculations; we compare the result for Si(111)1x1-As with that of the hydrogen-covered surface, whose calculation is shown by the full line in Fig. 2.6. It can be seen that the difference is quite small, but can be detected in a differential experiment. The difference is larger near 3, 4 and 5 eV, where surface transitions are more important.

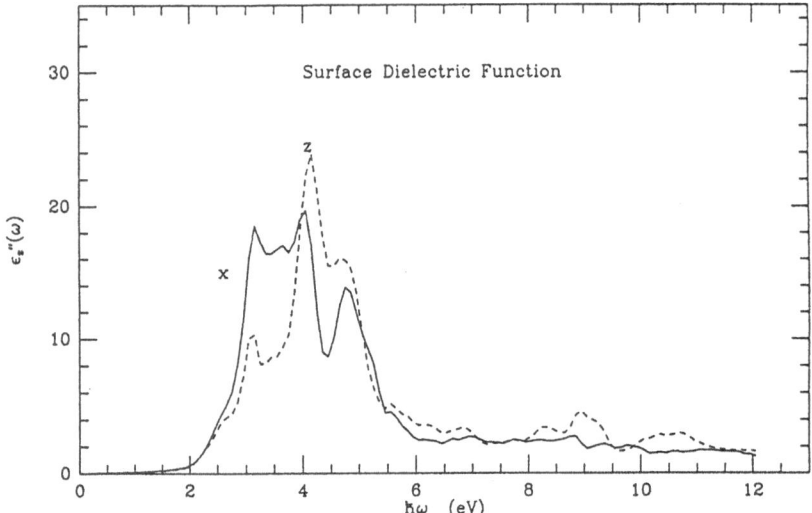

Fig. 2.6. Si(111) ellipsometry: the effective dielectric function. Full line: hydrogen-covered; dashed line: As-covered.

2.3.3 Calculations of the RAS Response from Si(001) Dimerised Surfaces

We believe that optical measurements carried out on single domain samples can resolve the controversy between two models of reconstruction, namely the symmetric versus the asymmetric dimer [38]. In Fig. 2.7, the RAS response

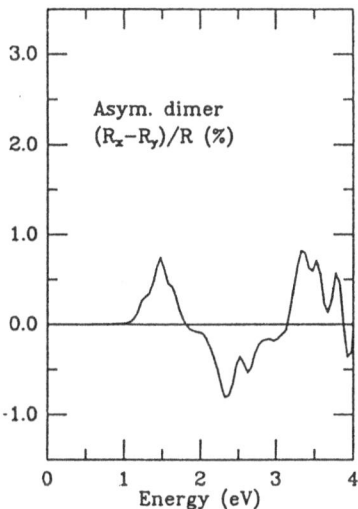

Fig. 2.7. Calculated RAS of Si(001) 2x1, for the asymmetric dimer model [39].

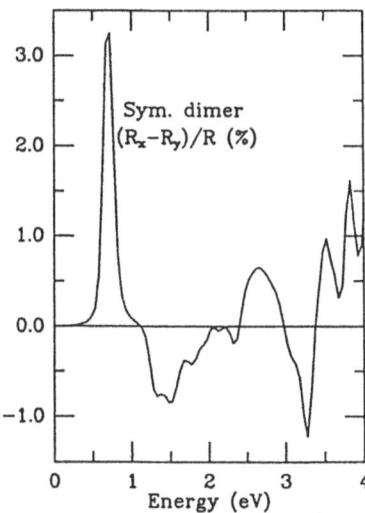

Fig. 2.8. Calculated RAS of Si(001) 2x1, for the symmetric dimer model [39].

calculated for the asymmetric dimer model is plotted. The x-direction is parallel to the dimers, while the y-direction is perpendicular to the dimer chains. The strong negative peak at 1.5 eV corresponds to transitions from the p-bond to the p-antibond within the same dimer [39]. The peak energy coincides with the mean separation of the surface bands based on these orbitals. The analogous calculation for the symmetric dimer is shown in Fig. 2.8. The energy of the peak mentioned above is less than 1 eV now, since the energy separation of the p-bond with respect to the p-antibond has decreased. The large difference between Figs. 2.7 and 2.8 leads us to conclude that performing RAS measurements on single domain samples might help to determine the real structure of this surface. However, the available SDR measurements, carried out using unpolarised light [13, 40], can also yield information on the structure. The calculated SDR spectrum for the asymmetric-dimer models, shown in Fig. 2.9, is indeed in very good agreement with the experimental spectrum; on the other hand, the SDR spectrum calculated assuming symmetric dimers, shown in Fig. 2.10, predicts the

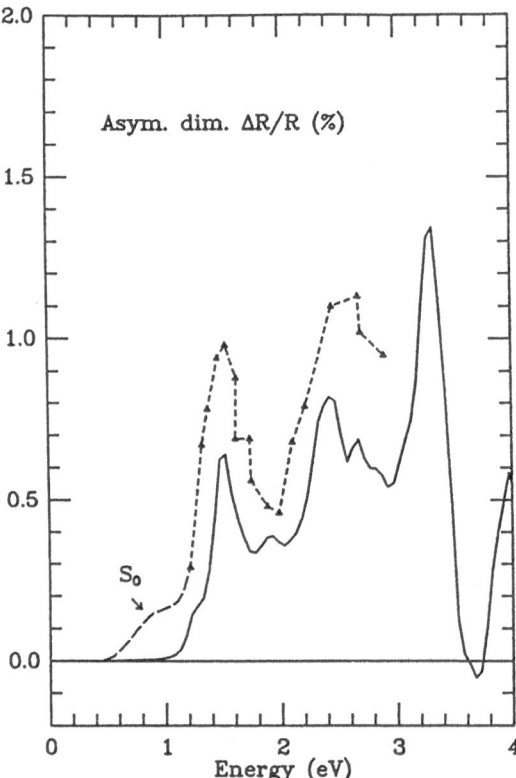

Fig. 2.9. Surface contribution to the reflectance of unpolarised light calculated for Si(001) 2x1 within the asymmetric dimer model [39].

strong peak at 0.7 eV which is not observed in the experiment. The weak S_0 structure could be due to the presence of a small number of symmetric dimers (less than 10%). Therefore, optical spectroscopy leads us to discard the possibility of a substantial presence of symmetric dimers on Si(001)2x1.

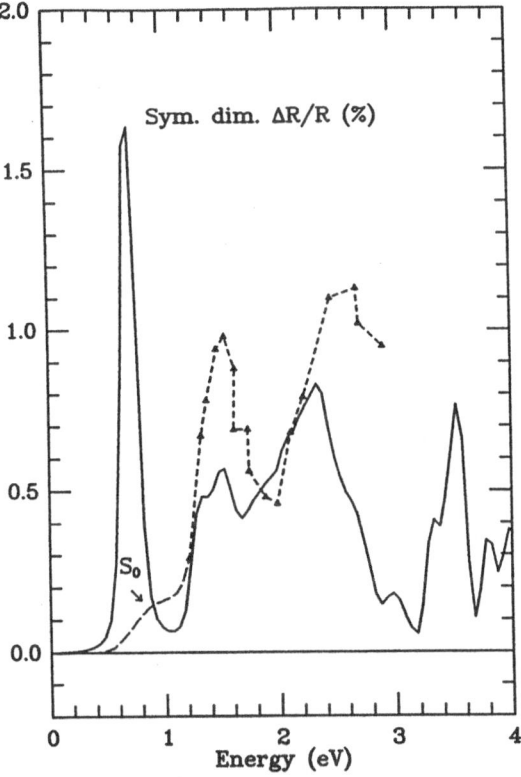

Fig. 2.10. Surface contribution to the reflectance of unpolarised light calculated for Si(001)2x1 within the symmetric dimer model [39].

2.3.4 Pseudopotential Calculations of GaP and GaAs Optical Properties

The self-consistent electronic structure determined using local pseudopotentials is expected to yield superior results compared to the tight-binding method. We use a plane-wave basis set and the local density approximation [27]. The computational effort is less in this case. The reason is that the momentum operator is diagonal in the plane-wave basis, but not in the local-orbital basis. A very accurate description of the ground state electronic properties of semiconductors can be obtained from first principles in the density functional (DF) scheme [41]. An intrinsic limitation of this approach occurs in the calculation of optical spectra: as is well known, DF theory fails to predict excited states [42], for which self-energy corrections are required [24]. The self-energy calculation is, however, very complicated and almost prohibitive in the case of low dimensional systems like surfaces, so that simplified approaches are required. The most straightforward way to avoid such a deficiency of DF calculations is to introduce an artificial enhancement of the exchange and correlation potential entering the Kohn-Sham equation, according to the X_α method [43]. This has been widely and successfully used in the calculation of the surface band structure of zincblende semiconductors [44] and we have applied it to the study of surface optical spectra.

The calculations we will report have been performed using the pseudopotential method. As in Ref. 44, we have chosen local pseudopotentials containing few parameters, adjusted to fit the energy levels of the free ions. These potentials have the advantage of converging rapidly enough in reciprocal space to allow an expansion of the wave functions in terms of a relatively small number of plane waves. The repeated slab method has been used in order to deal with the symmetry breaking introduced by the surface. The size of the unit cell is crucial in order to make a realistic comparison between theory and experiments. In particular, we have considered a unit cell made of 15 atomic layers + 8 empty layers corresponding to a size of about 4.5 nm along z. Such a large unit cell corresponds to a very large number of plane waves: about 2000 plane waves, using our rapidly converging local pseudopotentials and a kinetic energy cut-off of 6 Rydbergs. The same ingredients have been used for the calculation of the bulk band structure in order to allow meaningful comparison between bulk and surface energy spectra.

The eigenstates and eigenfunctions calculated for bulk and (110) surface terminated GaP and GaAs crystals are used to calculate the transition probabilities entering the imaginary part of the slab and bulk polarisabilities. The clean surfaces are assumed to be in the relaxed configuration [45]. In order to simulate the SDR spectrum, the calculations are repeated for unrelaxed H-covered surfaces. Transitions up to 1 Rydberg have been considered in the calculation of the imaginary parts in order to obtain realistic real parts of the dielectric functions by Kramers-Krönig transformation, in the energy range of interest (from 0 to approximately 6 eV). As an example, we show (Fig. 2.11) the calculated SDR spectrum for GaP(110), together with the experimental data. Good agreement between theory and experiment is found in this case.

The main result of the SDR and RAS calculations, carried out for GaAs and GaP (110) surfaces was to stress the importance of bulk states, perturbed by the surface, in determining RAS and SDR in the energy region of strong bulk absorption: we have found [27] that 90% of the RAS signal and 50% of the SDR signal originates from bulk states in this case. Of course, the signal arising from surface-perturbed bulk states is sensitive to surface chemistry and geometry, but not to the spectrum of surface states. Surface optical measurements achieve their full spectroscopic power only below the energy range of strong bulk absorption.

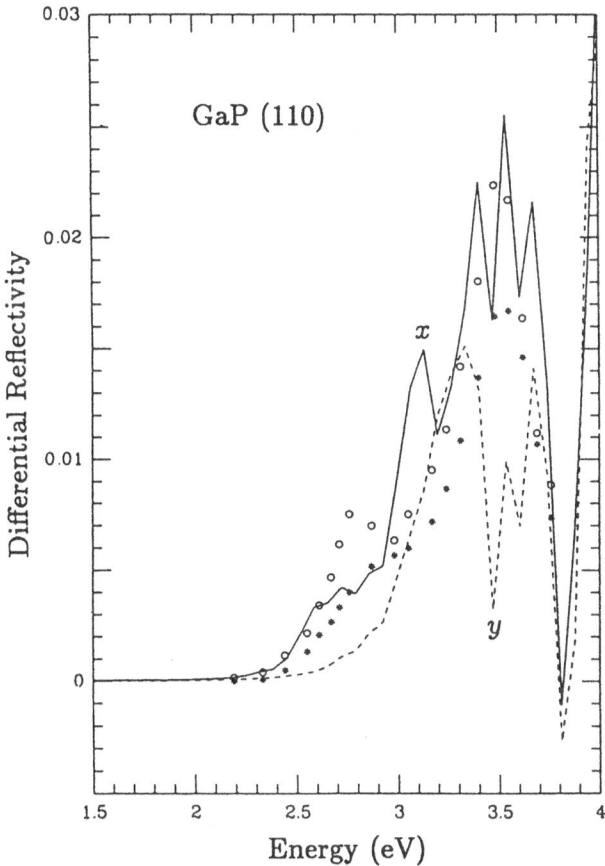

Fig. 2.11. SDR of GaP(110). Circles and stars: measured values for light polarised along [1$\bar{1}$0] (x) and [001] (y) respectively [14]. Solid and dashed lines: calculated $(R_{clean} - R_H)/R_H$ for H atoms aligned along the missing bonds [27].

2.4 The Bond Approach to the Linear Optical Response

Models for the lattice dynamics and optical properties of solids which treat a solid as an ordered array of polarisable atoms, ions or bonds have existed for some time. Early work by Born and Huang [46] and Maradudin *et al* [47] included electronic degrees of freedom, but focused on lattice dynamics. Later, Weber introduced 3 degrees of freedom for each pair of electrons in bonds in the bond charge model [48]. Quite recently, models of this type have successfully predicted phonon spectra of metals [49], alloys [50] and several phases of bulk Si [51].

Other workers have assumed that optical properties may be determined from long wavelength dipole wave modes of a crystal, where the electronic degrees of freedom are classical and sited on bonds, as in the bond charge model. In the 1970's, Litzman and Rósza [52] and Vlieger [53] considered propagation of radiation in media consisting of polarisable atoms. Around the same time Levine [54, 55] used analytic expressions to account for the linear and nonlinear susceptibilities of materials with a wide range of covalent to ionic character. Many recent calculations of surface optical properties have used bond models to calculate RAS [20], reflectance [56, 57] and SHG at surfaces [58]. The work to be described here has focused on calculating bond polarisabilities and hyperpolarisabilities, and how polarising bonds interact. This is an appropriate point to begin, since there are two major weaknesses of bond models for optical properties which have been used so far. Firstly, they have assumed that bonds in a solid interact as though the dipole moments induced in bonds were *point dipoles*, and secondly, they have relied on fitting bond polarisabilities to experiment for solids with one bond type only. The first assumption breaks down for near neighbour bonds, and the second restricts the model to solids consisting of a single bond type. The optical experiments which we wish to model involve much more subtle changes in the optical properties, such as surface reconstructions or chemisorption of gas molecules, etc., and so first principles calculations of the polarisabilities and bond interactions are required. We describe our method for calculating bond polarisabilities, etc., from first principles below.

2.4.1 First Principles Calculations of Bond Polarisabilities

Optical properties of solids, calculated using quantum mechanics, are usually expressed in terms of matrix elements of either the dipole or momentum operators and single-particle eigenvalues, where the basis states used to construct the matrix elements are *delocalised* over the entire solid. In order to calculate a *bond* polarisability using quantum mechanics, it is necessary to extract the response of *one* bond out of the whole system. This is achieved here using a wave function which naturally has orbitals which are localised in bonds. The wave functions which are used in this work are Generalised Valence Bond (GVB) wave functions. In this type of wave function, each electron is allocated its own spatial orbital which is permitted to overlap every other orbital to an extent which is determined in a self-consistent-field calculation. The wave function is a product of single-electron spatial functions and a complete set of spin eigenfunctions:

$$\Psi_{GVB} = A[\phi_{1A}\phi_{1B}\cdots\phi_{iA}\phi_{iB}\phi_{jA}\phi_{jB}\cdots\phi_{NA}\phi_{NB}]\Theta \qquad (2.19)$$

where A is the antisymmetrising operator, ϕ_{iA} and ϕ_{iB} are the pair of bond orbitals making up bond i, and Θ is a linear combination of spin eigenfunctions. The relationship between this wavefunction and the Hartree-Fock wavefunction and others is briefly described in Ref. 59. For computational convenience, two approximations are generally made to this wave function. These are the strong-orthogonality (SO) [60] and perfect-pairing (PP) approximations [60]. In the SO approximation, orbitals corresponding to different bonds ($i \neq j$ in equation 2.19) are constrained to be mutually orthogonal. Only orbitals within the same bond pair ($i = j$ in equation 2.19) have non-zero overlap now. In the PP approximation, the complete set of spin eigenfunctions is replaced by a single spin eigenfunction which may be written as a product of two-electron singlet spin pairings as

$$\Theta_{PP} = (\alpha\beta - \beta\alpha)(\alpha\beta - \beta\alpha)(\alpha\beta - \beta\alpha)... \tag{2.20}$$

Thus each *pair* of SOPP bond orbitals has entirely singlet spin-pairing. A further restriction on the SOPP-GVB wave function reduces it to a Hartree-Fock wave function ($\phi_{iA} = \phi_{iB}$) [58]. Examples of SOPP-GVB pairs for a Si-As bond and an As lone pair in a Si (111) surface are shown in Fig. 2.12 [61].

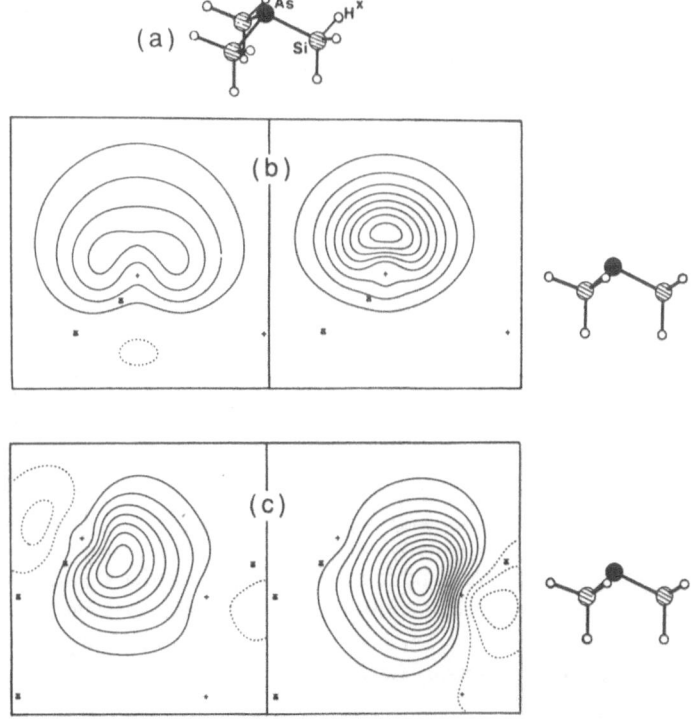

Fig. 2.12. (a) Schematic diagram of the $Si_3H_9^x$ cluster and contour plots of (b) As lone pair GVB orbitals and (c) the Si-As bond GVB orbitals. Contour intervals are 0.02 atomic units. Crosses indicate atomic positions in the plane while triangles and X symbols indicate atomic positions in front of and behind the plane, respectively [33, 61].

2.4.2 Bond Dipole Moments Expanded in the Local Field

Bond polarisabilities are calculated in one of two ways. *Either* all bonds in a cluster are permitted to relax from their zero-field orbital shapes in an electrostatic field [61], *or* bonds are allowed to relax individually or in pairs in an electrostatic field [62, 63]. In either case the individual bond dipole moments induced by the field are calculated as a function of field strength. Two neighbouring bonds polarised in an electric field strongly affect each other's bond dipole moment. This is referred to as interaction of the bonds. Obviously, in their unpolarised states, the electrons in the bonds have coulomb and exchange interactions, but the term bond interaction is reserved here for the way in which the dipole induced in one bond affects the dipole in an adjacent bond. Bond interactions are given a precise definition below.

Table 2.1. (a) Bare polarisabilities for Si-As, Si-Ga and C-H bonds at Si(111) and C(111) surfaces in units of $10^{-41}\ Cm^2V^{-1}$. (b) Dressed polarisabilities for Si-As and Si-Ga bonds at Si(111) and C(111) surfaces in units of $10^{-41}\ Cm^2V^{-1}$. The coordinate system has the $\parallel$ axis along the bond axis, the $\perp$ axis is perpendicular to the bond axis and in the same plane as the (111) surface normal, and the $\perp'$ axis is perpendicular to both the bond axis and the $\perp$ axis [after 61, 63].

Bond	(a) Bare polarisability				(b) Dressed polarisability		
		$\perp$	$\perp'$	$\parallel$		$\perp$	$\perp'$	$\parallel$
	$\perp$	5.4	0.0	0.6	$\perp$	5.7	0.0	1.2
Si-As	$\perp'$	0.0	6.2	0.0	$\perp'$	0.0	7.9	0.0
	$\parallel$	-0.3	0.0	13.3	$\parallel$	7.4	0.0	32.0
	$\perp$	10.2	0.0	4.4	$\perp$	5.1	0.0	-11.9
Si-Ga	$\perp'$	0.0	10.1	0.0	$\perp'$	0.0	15.0	0.0
	$\parallel$	2.1	0.0	14.9	$\parallel$	7.3	0.0	34.7
	$\perp$	4.0	0.0	0.0				
C-H	$\perp'$	0.0	4.0	0.0				
	$\parallel$	0.0	0.0	10.4				

One way of accounting for bond interactions is to consider a bond responding to the *local field* at the bond. For a particular bond, this is the field arising from all induced moments in all other bonds plus the applied field. Fitting the bond dipole moments, p_i, to the *local field*, E_j, using

$$p_i = \alpha_{ij}E_j + \beta_{ijk}E_jE_k + \dots \qquad (2.21)$$

yields the bond polarisability, α_{ij}, and first hyperpolarisability, β_{ijk}. The local field is calculated from first principles and varies strongly over the dimensions of the bond. It can actually exceed the strength of the applied field [61]. In the method we have developed, the local field used in equation 2.21 is the one obtained by averaging the local field, weighted by the bond charge density, over the bond. Bond polarisabilities for Si-As, Si-Ga and C-H bonds are given in Table 2.1(a). If the bond dipole moments are fitted to the *applied field*, instead of the local field when all bonds are allowed to relax together in a static field, unsymmetric polarisability matrices are obtained (Table 2.1(b)) which violate Onsager's principle [64]. Polarisabilities in Table 2.1(a) are almost symmetric and therefore they do not greatly violate Onsager's principle.

2.4.3 Bond Dipole Moments Expanded in the Applied Field

An expansion of the bond dipole moments in the *applied* field is given by the Taylor series [62]:

$$p_i = \frac{\partial p_i}{\partial F_j} F_j + \frac{1}{2} \frac{\partial^2 p_i}{\partial F_j \partial F_k} F_j F_k + ... \qquad (2.22)$$

where p_i is a component of a bond dipole moment and F_j is a component of the applied field at a bond site. In order to obtain the elements in the matrix $\partial p_i / \partial F_j$ $(= P_{ij})$, it is necessary to be able to selectively turn on the applied field at individual bonds. This is achieved by freezing all but the selected bond(s) in their ground state orbitals. If the indices i and j coincide on a particular bond, the elements of P_{ij} are a 3x3 polarisability matrix for that bond. If the indices are on different bonds, the elements of P_{ij} are a 3x3 bond interaction matrix. This gives a precise definition to the term bond interaction introduced above. An expansion of this kind has been calculated for n-hexane [61] in order to compare results of *ab initio* calculations of P_{ij} to the point dipole interaction model mentioned in the introduction. That is, a model in which bonds are treated as point-like polarisable entities at bond midpoints, which respond to the local field at that site. Since the dipoles induced in bonds are point dipoles, the local field, E_i, at the ith bond site is:

$$E_i = F_i + \frac{3r_i(r_k p_k') - r^2 p_i'}{4\pi\varepsilon_0 r^5} \qquad (2.23)$$

F_i is the applied field, r is the magnitude of the vector (with components r_i) connecting the two bonds, and the induced dipole on the second bond has components p_i'. The induced dipole, bond polarisability and applied field at the neighbouring bond are indicated with a prime to distinguish them from the original bond. Equating the bond dipole moment at site i to the product of the polarisability and the local field at that site permits partial derivatives with respect to the applied field at either site to be obtained. Differentiation of the dipole in equation 2.22 with respect to applied field leads to the bond polarisability, α_{ij}, when the applied field, F_j, is at the same site as the induced dipole:

$$\alpha_{ij} = \frac{\partial p_i}{\partial F_j} \tag{2.24a}$$

and the 3x3 point dipole interaction matrix in equation 2.24b, when the applied field, F_j, is at a neighbouring bond:

$$\frac{\partial p_i}{\partial F'_j} = \alpha_{ij} \frac{3r_j(r_k \alpha'_{kj}) - r^2 \delta_{kj} \alpha'_{kj}}{4\pi\varepsilon_0 r^5} \tag{2.24b}$$

Equation 2.24b can be written as:

$$\frac{\partial p_i}{\partial F'_j} = \alpha_{ij} f_{jk} \alpha'_{kj} \tag{2.25}$$

where f is the transfer tensor given by:

$$f_{jk} = \frac{3r_j r_k - r^2 \delta_{jk}}{4\pi\varepsilon_0 r^5} \tag{2.26}$$

We find that the point dipole approximation grossly overestimates (by a factor of 2 to 10 times) the bond interaction strengths between nearest neighbour bonds, but it is a good approximation for bonds which are 3rd or 4th nearest neighbours [62].

References

1.   Aspnes, D.E., Studna, A.A.: Physical Review Letters *54*, 1956 (1985)
2.   Aspnes, D.E.: Journal of Vacuum Science and Technology, *B 3*, 1138 (1985). *Ibidem B 3*, 1498 (1985)
3.   Chang, Y.C., Aspnes, D.E.: Physical Review *B 41*, 12002 (1990)
4.   Kamiya, I., Aspnes, D.E., Tanaka, H., Florez, L.T., Harbison, J.P., Bhat, R.: Physical Review Letters *68*, 627 (1992)
5.   Acosta-Ortiz, S.E., Lastras-Martinez, A.: Solid State Communications *64*, 809 (1987)
6.   Berkovits, V.L., Makarenko, I.V., Minashvili, T.A., Safarov, V.I.: Solid State Communications *56*, 449 (1985)
7.   Berkovits, V.L., Ivantsov, L.F., Makarenko, I.V., Minashvili, T.A., Safarov, V.I.: Solid State Communications *64*, 767 (1987)

8. Chiarotti, G., Del Signore, G., Nannarone, S.: Physical Review Letters *21*, 1170 (1968)
9. Chiarotti, G., Nannarone, S., Pastore, R., Chiaradia, P.: Physical Review *B 4*, 3398 (1971)
10. Chiaradia, P., Chiarotti, G., Nannarone, S., Sassaroli, P.: Solid State Communications *26*, 813 (1978). Nannarone, S., Chiaradia, P., Ciccacci, F., Memeo, R., Sassaroli, P., Selci, S., Chiarotti, G.: *ibidem 33*, 593 (1980)
11. Chiaradia, P., Cricenti, A., Selci, S., Chiarotti, G.: Physical Review Letters *52*, 1145 (1984)
12. Selci, S., Chiaradia, P., Ciccacci, F., Cricenti, A., Sparvieri, N., Chiarotti, G.: Physical Review *B31*, 4096 (1985)
13. Wierenga, P.E., van Silfhout, A., Spaarnay, M.J.: Surface Science *87*, 43 (1979). *Ibidem 99*, 59 (1980)
14. Selci, S., Ciccacci, F., Cricenti, A., Felici, A.C., Goletti, C., Chiaradia, P.: Solid State Communications *62*, 833 (1987)
15. Del Sole, R., Selloni, A.: Solid State Communications *50*, 825 (1984)
16. McIntyre, J.D.E., Aspnes, D.E.: Surface Science *24*, 417 (1971)
17. Philpott, M.R.: Advances in Chemical Physics *23*, 227 (1973)
18. Munn, R.W.: Molecular Physics *64*, 1 (1988). Munn, R.W.: Journal of Chemical Physics *97*, 4532 (1992)
19. Mochan, W.L., Barrera, R.: Physical Review Letters *55*, 1192 (1985)
20. Wijers, C.M.J., Poppe, G.P.M.: Physical Review *B 46*, 7605 (1992)
21. Feibelman, P.J.: Progress in Surface Science *12*, 287 (1982)
22. Bagchi, A., Barrera, R.G., Rajagopal, A.K.: Physical Review *B 20*, 4824 (1979)
23. Ehrenreich, H. in: Tauc, J. (ed.) The Optical Properties of Solids. Academic Press, New York 1966
24. Hedin, L.: Physical Review *139*, A796 (1965). Hedin, L., Lundquist, S.: Solid State Physics *23*, 1 (1969). Hybertsen, M.S., Louie, S.G.: Physical Review *B 34*, 5390 (1986). Godby, R.W., Schluter, M., Sham, L.J.: Physical Review *B 37*, 10159 (1988)
25. Del Sole, R.: Solid State Communications *37*, 537 (1981)
26. Del Sole, R., Fiorino, E.: Physical Review *B 29*, 4631 (1984)
27. Manghi, F., Del Sole, R., Selloni, A., Molinari, E.: Physical Review *B 41*, 9935 (1990)
28. Del Sole, R., Fiorino, E.: Journal of the Physical Society of Japan *49 supplement A*, 1133 (1980)
29. Vogl, P., Hjalmarson, H.P., Dow, J.D.: Journal of Physics and Chemistry of Solids *44*, 365 (1983)
30. Selloni, A., Marsella, P., Del Sole, R.: Physical Review *B 33*, 8885 (1986)
31. Pandey, K.C.: Physical Review Letters *47*, 1913 (1981). *Ibidem 49*, 223 (1982)
32. Patterson, C.H., Messmer, R.P.: Physical Review *B 39*, 1372 (1989)
33. Hybertsen, M.S., Louie, S.G.: Physical Review *B 38*, 4033 (1988)
34. Uhrberg, R.I.G., Bringans, R.D., Olmstead, M.A., Bachrach, R., Northrup, J.E.: Physical Review *B 35*, 3945 (1987)
35. Kelly, P.V., Tang, Z.-R., Woolf, D.A., Williams, R.H., McGilp, J.F.: Surface Science *251/252*, 87 (1991)
36. Rossow, U., Frotscher, U., Richter, W., Zahn, D.R.T.: Surface Science *287/288*, 718 (1993)
37. Reining, L., Del Sole, R., Cini, M., Ping, J.G.: Physical Review *B 50*, 8411 (1994)
38. Badziag, P., Verwoerd, W.S., Van Hove, M.A.: Physical Review *B 43*, 2058 (1991). Tsuda, M., Hoshino, T., Oikawa, S., Ohdomari, I.: Physical Review *B 44*, 11241 (1991). *Ibidem B 44*, 11248 (1991)
39. Shkrebtii, A.I., Del Sole, R.: Physical Review Letters *70*, 2645 (1993)

40. Chabal, Y.J., Christman, S.B., Chaban, E.E., Yin, M.T.: Journal of Vacuum Science and Technology, *A 1*, 1241 (1983)
41. Pickett, W.E.: Comments on Solid State Physics *12*, 1 (1985)
42. Pickett, W.E.: Comments on Solid State Physics *12*, 57 (1985)
43. Slater, J.C.: Journal of Chemical Physics *43*, S228 (1965)
44. Manghi, F., Bertoni, C.M., Calandra, C., Molinari, E.: Physical Review *B 24*, 6029 (1981)
45. Kahn, A.: Surface Science Reports *3*, 193 (1983)
46. Born, M., Huang, K.: Dynamical Theory of Crystal Lattices. Clarendon, Oxford 1954
47. Maradudin, A.A., Montroll, E.W., Weiss, G.H, Ipatova, I.P.: Theory of Lattice Dynamics in the Harmonic Approximation. Academic, New York 1971
48. Weber, W.: Physical Review Letters *33*, 371 (1974). Weber, W.: Physical Review *B 15*, 4789 (1977)
49. Li, M., Goddard, W.A.: Physical Review *B 40*, 12155 (1989)
50. Schultz, P.A., Messmer, R.P.: Physical Review *B 45*, 7467 (1992)
51. Wang, H.-X.: PhD dissertation, University of Pennsylvania, Pennsylvania 1992
52. Litzman, O., Rózsa, P.: Surface Science *66*, 542 (1977)
53. Vlieger, J.: Physica *64*, 63 (1973)
54. Levine, B.F.: Physical Review Letters *22*, 787 (1969). *Ibidem 25*, 440 (1970). *Ibidem 33*, 368 (1974)
55. Levine, B.F.: Physical Review *B 7*, 2591 (1973). *Ibidem B 7*, 2600 (1973)
56. Wijers, C.M.J., Del Sole, R., Manghi, F.: Physical Review *B 44*, 1825 (1991)
57. Chen, W., Schaich, W.L.: Surface Science *218*, 580 (1990)
58. Schaich, W.L., Mendoza, B.S.: Physical Review *B 45*, 14279 (1992)
59. Patterson, C.H., Messmer, R.P.: Physical Review *B 42*, 7530 (1990)
60. The SO and PP approximations were first suggested in: Hurley, A.C., Lennard-Jones, J.E., Pople, J.A.: Proceedings of the Royal Society (London) Series A *220*, 446 (1953)
61. Patterson, C.H., McGilp, J.F., Weaire, D.: Journal of Physics: Condensed Matter *4*, 4017 (1992)
62. Patterson, C.H.: Chemical Physics Letters *213*, 59 (1993)
63. Patterson, C.H.: Surface Science *304*, 365 (1994)
64. Nye, J.F.: Physical Properties of Crystals. Clarendon, Oxford 1967

Chapter 3. Spectroscopic Ellipsometry

Uwe Rossow

Institut für Festkörperphysik, Technische Universität, D-10623 Berlin

3.1 Introduction

Ellipsometry dates back to the middle of the last century [1, 2], and it has been used since then for the determination of the optical properties of metals, semiconductors and insulators [3]. Ellipsometry is nowadays quite intensively employed in semiconductor characterisation and has the potential for *in situ* diagnostics of surfaces. The principle of ellipsometry is based on the fact that the status of light polarisation is changed when light is reflected from a surface. This change can be related to the dielectric function of the reflecting material, as discussed in Sect. 1.3.

In the most common experimental configuration, linear polarised light is incident on the surface and the elliptical polarisation status of the reflected light (Fig. 3.1) is analysed. As discussed in Sect. 1.3, the complex ratio of the reflection coefficients, r_p and r_s of light polarised parallel (p-) and perpendicular (s-) to the plane of incidence, defined by

$$\rho = r_p / r_s \tag{3.1}$$

can be derived from the polarisation status before and after reflection. For historical reasons the ratio is often expressed by the ellipsometric angles, Δ and Ψ:

$$\rho = r_p / r_s = \tan \Psi \exp i\Delta \tag{1.14}$$

Fresnel equations relate this equation to the bulk complex dielectric function, ε. If the sample is isotropic and homogeneous, ε depends only on the photon energy, $E = \hbar\omega$, of the incident light. Quite often the material is not homogeneous but consists of layers or is otherwise structured. In such a case the dielectric function determined by ellipsometry is an average over the region penetrated by the incident light and is called the *pseudodielectric function* or, in the following, the *effective dielectric function*, and written as $\langle\varepsilon\rangle$. If the sample structure is not too complicated, the effective dielectric function can be simulated by appropriate models like the three-phase (vacuum-layer-substrate) model. In these cases, layer and substrate properties can be separated and layer properties such as the layer dielectric function or the layer thickness can be determined.

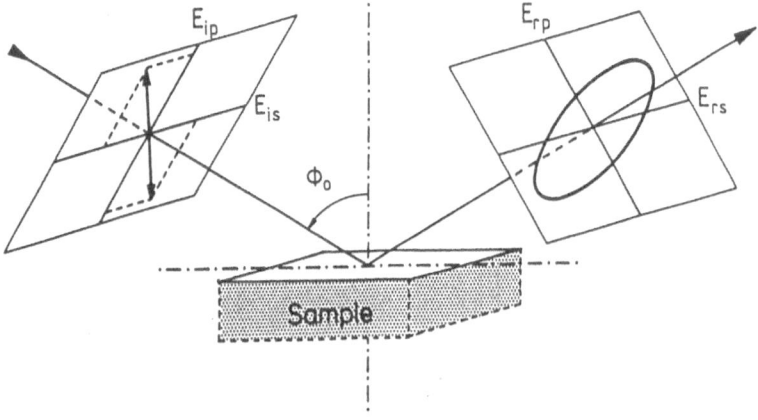

Fig. 3.1. The principle of measurement. Linear polarised light is incident on the sample at an angle ϕ_0, and the elliptically polarised reflected light is detected.

Quite often it is not possible to separate layer and substrate properties, because either the geometry or the dielectric function of the layer(s) is not known. In such a case, the existence of surface layers may still be clearly observable in $\langle\varepsilon\rangle$, and the modifications of the surface layer monitored very sensitively (in the submonolayer range). This is one of the main aspects of ellipsometry for *in situ* diagnostics of surfaces or interfaces. Moreover, the ellipsometric measurement can be done quite quickly and thus real time growth control is possible.

This chapter deals with applications of spectroscopic ellipsometry (SE) in the analysis of surfaces and interfaces of semiconductors. The focus will be on the visible and UV spectral range, where electronic properties are probed. Applications of ellipsometry in the infra-red (IR) for investigations of vibrational properties can be found elsewhere [4-7]. The chapter is organised as follows. In Sect. 3.2, the principle of the method and experimental details are given. The interpretation of $\langle\varepsilon\rangle$ is described in the Sect. 3.3, and selected examples are explained in detail in Sect. 3.4.

3.2 The Method

Ellipsometers differ mainly by the way the polarisation of the reflected light is measured. A detailed description of the various experimental configurations can be found in [8]. Photometric ellipsometers, which measure the intensity of the reflected light while modulating the polarisation of the incident or the reflected light, are now widely used.

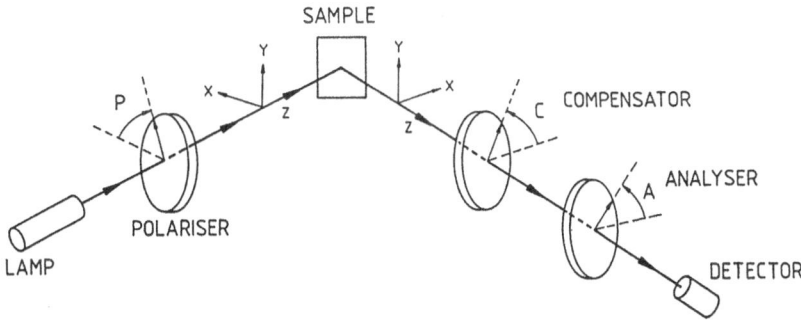

Fig. 3.2. A schematic set-up of an ellipsometer, where L is the light source, P the polariser, S the sample, C the compensator, A the analyser, and D the detector.

A schematic set-up of an ellipsometer is shown in Fig. 3.2, consisting of a light source, L, a polariser, P, an optional compensator, C, the sample, S, the analyser, A, and a detector, D. Following the notation given in [8], P, C, and A describe both the optical elements and their angles relative to the plane of incidence. The compensator is a phase shifting element (usually a quarter-wavelength plate) and is necessary for null ellipsometry. This configuration is called PSCA (polariser-sample-compensator-analyser), which is equivalent to a PCSA arrangement. All commercially available ellipsometers, and nearly all of the ellipsometers used for the visible and near UV spectral region, have this configuration. They differ only in the ways in which the incident light is polarised, and the polarisation state of the reflected light is determined.

3.2.1 Null Ellipsometry

The oldest type of ellipsometer is the *null ellipsometer*. In null ellipsometry, the angles of two of the three elements, P, C, and A are varied such that the intensity measured by the detector is minimised (null-intensity condition) but, in practice, much more complicated procedures are used to find the angles P, C, and A of a null-intensity condition [9]. The earlier null ellipsometers were manually operated, and were both slow and difficult to use. Automising the null ellipsometer is possible by installing Faraday rotators behind the polarisers, or by attaching motors to the components [8, 10]. The Faraday cells limit the useful spectral range to the visible, so that these ellipsometers are not useful for semiconductors with optical gaps in the near UV spectral range, like Si and GaAs. Another problem is that the phase shift of the compensator varies with the wavelength used, and this has to be corrected during the measurement. These disadvantages made ellipsometry less attractive in the past.

3.2.2 Photometric Ellipsometry

Photometric ellipsometry became possible with the development of inexpensive computers and analogue-to-digital converters [11-15]. These ellipsometers can be easily computer controlled and consist of simple elements. Most spectra referenced in this chapter were recorded using spectroscopic photometric ellipsometers. This type of ellipsometer is commercially available. In contrast to null ellipsometry, at least one of the optical elements modulates the polarisation state in photometric ellipsometry. The variation of the light intensity caused by this modulation at the detector is measured. Two types of modulation can be distinguished. Firstly, the polariser, (or analyser or compensator) can be rotated and the ellipsometer is then called a rotating polariser ellipsometer, etc. Secondly, the phase shift of the compensator may be varied with time, usually with a photoelastic modulator (PEM). PEMs exploit a strain-induced birefringence in quartz, as discussed below.

The most common type of a photometric ellipsometer is, however, the rotating analyser ellipsometer [11-13], illustrated in Fig. 3.3. The light sources used are Xe high pressure arc-lamps and halogen lamps. A Xe high pressure arc-lamp has a continuous radiation spectrum between 1.5 eV (830 nm) and 6.2 eV (200 nm), while the useful spectral range of a halogen lamp is from the near infrared (around 1 eV) region up to approximately 4 eV (310 nm) [16]. Since Xe has spectral lines with very high intensity below 1.5 eV [16] and a halogen lamp has a smooth characteristic, the latter is preferred in the near-IR spectral region. Halogen lamps are, however, limited in intensity in the UV region above 4 eV, and Xe arc lamps are mostly used for investigations in the visible and UV spectral region.

The polariser and analyser are usually Rochon or Glan(-air) prisms [17]. Both types are made from the (uniaxial) birefringent materials quartz or calcite. Since the refractive index of light polarised parallel to that axis differs from that for light polarised perpendicular to the axis, one can produce a beam which passes the prism unaffected and is linear polarised, and one which is deflected and polarised perpendicular to the former one. The transmission axis angle, P, of the polariser is usually fixed when taking a spectrum. Although not critical, a value of 30° for P has been found to optimise the precision for a wide range of substrate materials [11, 18].

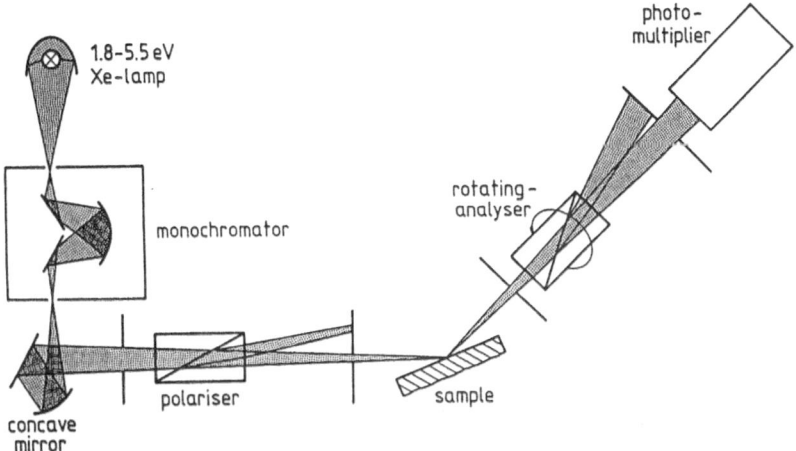

Fig. 3.3. A typical rotating analyser ellipsometer with a Xe-lamp as light source, Rochon prisms as polariser and analyser, and a photomultiplier as detector. The arc of the Xe lamp is imaged on the entrance slit of a monochromator. The monochromatic light passes the polariser and is focused on the sample. The reflected light is then analysed by a second polariser (the rotating analyser) and finally detected by a photomultiplier.

The analyser rotates at a constant frequency between 25 and 60 Hz. The rotation frequency is limited by vibrations which originate from mechanical imperfections. The angle of the transmission axis of the analyser is measured by an angle encoder, or resolver, mounted on the holder of the analyser. An angle of incidence, ϕ_0, between 60° and 80° provides highest sensitivity to film properties for semiconductor substrates [8, 12]. For Si, the pseudo-Brewster condition (where the reflectance of p-polarised light has its minimum) is around 75.6° for a wavelength of 632.8 nm.

Until recently, photomultipliers were mostly used as detectors. The advantage of the photomultiplier is the high quantum efficiency in the visible and near-UV spectral region, the high gain providing high output currents (typically several μA), and the extremely linear relationship between light intensity and signal. An alternative is a Si diode, which is inexpensive and now has an adequate sensitivity for many applications. Recently, optical multichannel analyser (OMA) systems have been based on Si diode arrays [19].

For absorbing materials, the reflected light is elliptically polarised. The shape of this ellipse is transferred into a sinusoidal varying intensity by the rotation of the analyser. The signal of the detector is at maximum (minimum) when the transmission axis of the analyser is parallel (perpendicular) to the major axis of the ellipse (Fig. 3.4).

Another possibility is to modulate the polarisation state by using a compensator with a time-varying phase shift, such as the PEM mentioned above. The stress induced birefringence in quartz is used to modulate the phase shift of the polarised light in two perpendicular directions [20]. This method is becoming more popular because the set-up needs no special mechanical equipment, in contrast to rotating analyser configurations [11, 21]. This type of modulation is inherently faster than that of the rotating analyser, because the modulating frequency of the PEM may be typically 50 kHz. The disadvantage of a PEM is that the amplitude of its phase shift has to be known, and this varies with both wavelength and mechanical imperfections in the modulator. More sophisticated signal processing is also required.

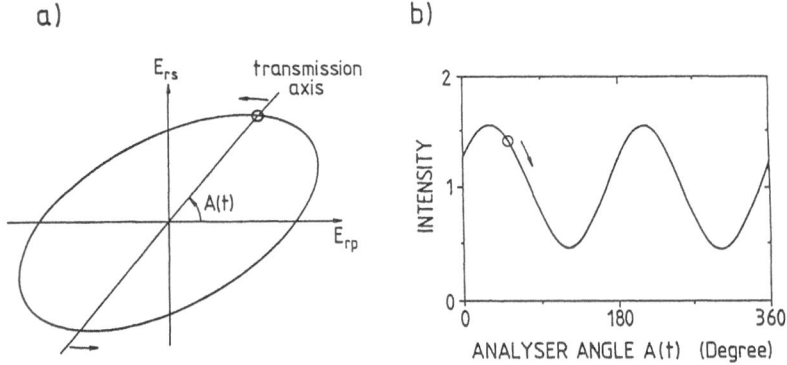

Fig. 3.4. Analysis of the ellipse. The signal of the detector varies depending on the relative orientation of analyser transmission axis and ellipse.

3.2.3 Measuring $<\varepsilon>$

The intensity is modulated in all types of photometric ellipsometer. The knowledge of the time dependence of the signals allows the determination of the polarisation state of the reflected light. For the rotating analyser type, the intensity at the detector can be expressed as a function of the time-varying angular position, A, of the transmission axis of the analyser [11]:

$$I(A) = I_0 (1 + a_2 \cos 2A + b_2 \sin 2A) \tag{3.2}$$

The polarisation state of the reflected light can be determined from the coefficients, a_2 and b_2. These coefficients contain the information on the optical response of the sample and may be simply obtained by a Fourier transform of the signal. It is important to note that these Fourier coefficients are calculated from relative intensities and thus, in principle, photometric ellipsometry is not sensitive to variations of the absolute intensity of the lamp. The ratio of the reflection coefficients, ρ, defined in equation 3.1, can be calculated from a_2 and b_2 [11].

The effective dielectric function of the system, $<\varepsilon>$, is related to ρ and the angle of incidence, ϕ_0, by [8]:

$$<\varepsilon>= \sin^2 \phi_0 + \sin^2 \phi_0 \tan^2 \phi_0 \left[\frac{1-\rho}{1+\rho}\right]^2 \tag{3.3}$$

As $<\varepsilon>$ is a complex quantity, two unknowns of the sample affecting the optical response can be determined for each wavelength.

The brackets around $<\varepsilon>$ are a reminder that this dielectric function is obtained under the assumption it contains only the bulk response from a semi-infinite, isotropic and homogeneous solid. These assumptions are not fulfilled for samples composed of more than one phase (layer on substrate). Even for cubic semiconductors (which are isotropic), surface roughness, contamination, or reconstruction has to be considered. Therefore, it is an effective dielectric function which is obtained in all cases.

3.2.4 Trends and New Developments

There have been many new ellipsometer types developed in the last few years [8, 10, 22-24]. Efforts at present are, however, concentrating on an extension of the spectral range and achieving better time resolution. The IR spectral range, probing vibrational properties, is now accessible [4, 5, 7], as well as the VUV (vacuum UV [24, 25]), where higher energy optical gaps and plasmons can be observed. Time resolution reached is sufficient for growth control at a fixed spectral position [26]. It is further improved by applying PEM, as mentioned above, and OMAs [19]. The best time resolution reported for a single wavelength ellipsometer, based on a interferometer, is 25 ns [27].

Another important development is the combination of spectroscopic ellipsometry with varying angle of incidence (VASE) [28]. For anisotropic samples and multilayer films, varying the angle of incidence provides more data for analysis of film properties. It is also possible to identify the contribution of depolarisation to the optical response. The problems of varying the angle of incidence are that the spot size and, for non-ideal alignment, the location of the spot, varies. Therefore, the samples must be laterally homogeneous.

3.3 Interpretation of the Effective Dielectric Function

In this section the evaluation and interpretation of $<\varepsilon>$ in the most common situations is discussed. The application of these methods for some selected examples is given in the next section.

3.3.1 Lineshape Analysis of Optical Gaps

Figure 3.5 shows the real and imaginary part of the dielectric function of GaAs [3]. Four major structures are visible in the dielectric function, namely E_1, $E_1 + \Delta_1$, E_0', and E_2. They are van Hove singularities of the joint density of states (JDOS), often referred to as interband critical points or optical gaps [29-31]. The interband critical points are shown in the bandstructure in Fig. 3.6.

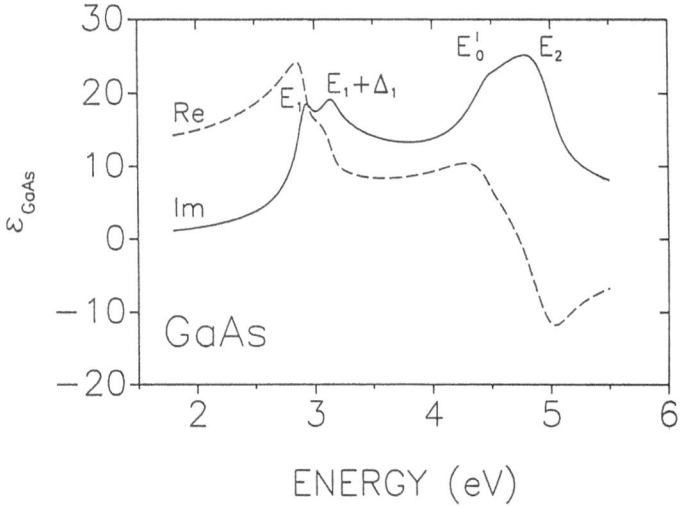

Fig. 3.5. The dielectric function of GaAs (after [3]). The structures labelled E_1, $E_1 + \Delta_1$, E_0', and E_2 are the interband critical points of GaAs in the visible and UV spectral region.

The spectral position and broadening of the optical gaps were mostly determined historically by modulation techniques like photoreflectance and electroreflectance [30, 33]. In these investigations it was found, empirically, that the lineshape of the optical transitions can be described using functions of the form [30, 31, 33]:

$$<\varepsilon>'' = Ae^{i\phi}(E - E_g + i\Gamma)^{-n} \qquad (3.4)$$

where ϕ and n are listed in Table 3.1, and A is the amplitude, E_g the gap energy, and Γ the broadening parameter of the optical transition. The second derivative with respect to energy, $<\varepsilon>''$, is chosen instead of $<\varepsilon>$ for the fitting procedure in order to reduce the influence of artefacts like oxide overlayers. The exponent, n, depends on the dimensionality of the optical transition. The angle, ϕ, was introduced to enable all functions describing the lineshape for the various kinds of optical gaps to be written in one form [30, 31].

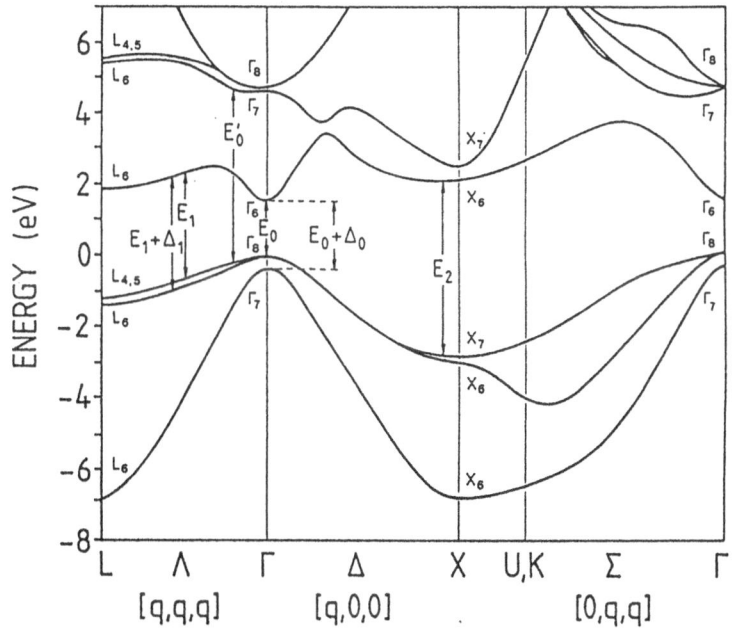

Fig. 3.6. The bandstructure of GaAs. The interband critical points E_1, $E_1 + \Delta_1$, E_0', and E_2 of GaAs in the visible and UV spectral region are shown (after [32]).

For values of ϕ differing from $0°$, $90°$, $180°$, and $270°$, a mixing between Re $<\varepsilon>$ and Im $<\varepsilon>$ is occurring, which may be due, for example, to many-body contributions to $<\varepsilon>$ [31, 34]. For GaAs and many other semiconductors, $<\varepsilon>$ is modified by the electron-hole interaction (excitonic effects). This interaction drastically increase the JDOS near the gaps, leading to sharper structures and higher values of Im $<\varepsilon>$. The optical gaps E_0, E_1, and $E_1 + \Delta_1$ of GaAs are dominated by these excitonic effects [30, 31, 35, 36]. Such excitonic transitions can also be included in equation 3.4 by using $n = 3$ [30, 31, 33]. The E_0' and E_2 transitions of GaAs are examples of two dimensional (2D) transitions.

The energetic positions of the optical gaps depend on the bandstructure, and are thus a characteristic of the material. Therefore, the chemical composition of a substrate or thick layer can be determined from the energetic position of the gaps [37-39]. In the case of microcrystallites, the energetic positions and broadening parameters of gaps depend on the crystallite size [40, 41], allowing the (dominant) crystallite size in a nanocrystalline layer to be determined, after calibration.

Table 3.1. All possible combinations of ϕ and n for the van-Hove singularities of the joint density of states

Dimensionality	Type of Singularity	ϕ	n
3D	Minimum	0°	3/2
	Saddle point	90°	
	Saddle point	180°	
	Maximum	270°	
2D	Minimum	0°	2
	Saddle point	90°	
	Maximum	180°	
1D	Minimum	270°	5/2
	Maximum	0°	
Excitonic			3

3.3.2 A Single Layer on a Substrate

In order to determine the optical properties of a thin layer (which may be the surface layer) on a substrate, the contribution of the substrate to $<\varepsilon>$ has to be evaluated. This can be done using the *three-phase model*. Comparison of the calculated and measured $<\varepsilon>$ values allows the determination of both ε_{layer} and d_{layer} (the layer thickness), as discussed below.

3.3.2.1 The Three-Phase Model

This model assumes that the ambient-layer and layer-substrate interfaces are abrupt, with properties changing discontinuously. Furthermore, the substrate is assumed to be absorbing, with no light being reflected back from the reverse side of the sample. This model can easily be extended to the more general case of a layered structure using the Abeles formalism [8].

The ratio, ρ, for an *isotropic* layer (1) on an *isotropic* substrate (2) in the ambient (0), for an angle of incidence, ϕ_0, is given by [8]:

$$\rho = \frac{r_{01p} + r_{12p}e^{+i2\beta}}{1 + r_{01p}r_{12p}e^{+i2\beta}} \frac{1 + r_{01s}r_{12s}e^{+i2\beta}}{r_{01s} + r_{12s}e^{+i2\beta}} \tag{3.5}$$

where r_{ijp}, r_{ijs} are the reflection coefficients for p- and s-polarised light, respectively, at the interfaces from phase i =0,1 to phase j = 1,2. These Fresnel coefficients are:

$$r_{01p} = \frac{\varepsilon_1 \sqrt{\varepsilon_0 - \varepsilon_0 \sin^2 \phi_0} - \varepsilon_0 \sqrt{\varepsilon_1 - \varepsilon_0 \sin^2 \phi_0}}{\varepsilon_1 \sqrt{\varepsilon_0 - \varepsilon_0 \sin^2 \phi_0} + \varepsilon_0 \sqrt{\varepsilon_1 - \varepsilon_0 \sin^2 \phi_0}} \qquad (3.6)$$

$$r_{12p} = \frac{\varepsilon_2 \sqrt{\varepsilon_1 - \varepsilon_0 \sin^2 \phi_0} - \varepsilon_1 \sqrt{\varepsilon_2 - \varepsilon_0 \sin^2 \phi_0}}{\varepsilon_2 \sqrt{\varepsilon_1 - \varepsilon_0 \sin^2 \phi_0} + \varepsilon_1 \sqrt{\varepsilon_2 - \varepsilon_0 \sin^2 \phi_0}} \qquad (3.7)$$

$$r_{01s} = \frac{\sqrt{\varepsilon_0 - \varepsilon_0 \sin^2 \phi_0} - \sqrt{\varepsilon_1 - \varepsilon_0 \sin^2 \phi_0}}{\sqrt{\varepsilon_0 - \varepsilon_0 \sin^2 \phi_0} + \sqrt{\varepsilon_1 - \varepsilon_0 \sin^2 \phi_0}} \qquad (3.8)$$

$$r_{12s} = \frac{\sqrt{\varepsilon_1 - \varepsilon_0 \sin^2 \phi_0} - \sqrt{\varepsilon_2 - \varepsilon_0 \sin^2 \phi_0}}{\sqrt{\varepsilon_1 - \varepsilon_0 \sin^2 \phi_0} + \sqrt{\varepsilon_2 - \varepsilon_0 \sin^2 \phi_0}} \qquad (3.9)$$

where ε_0, $\varepsilon_1 (= \varepsilon_{layer})$ and ε_2 are the complex dielectric functions of ambient, layer and substrate, respectively. Note the similarity to equation 1.10, where $\varepsilon_0 = 1$ for vacuum. The phase shift, β (see equation 1.8), is given by :

$$\beta = (2\pi d_{layer} / \lambda) \sqrt{\varepsilon_{layer} - \varepsilon_0 \sin^2 \phi_0} \qquad (3.10)$$

where λ is the wavelength of EM radiation in vacuum. A calculated effective dielectric function, $<\varepsilon>_{calc}$, can now be found from equation 3.3, using equation 3.5 to determine ρ, and the result compared with $<\varepsilon>$, the experimental value.

3.3.2.2 Determination of the Layer Properties

Different possibilities may occur in the analysis of a layer on a substrate. The easiest task is to determine the unknown thickness, d_{layer}, of the layer with known substrate and layer dielectric function. The layer thickness, d_{layer}, is varied such that the function

$$\Xi = \sum_{\hbar\omega} |<\varepsilon> - <\varepsilon>_{calc}|^2 \qquad (3.11)$$

(or equivalent definitions using ρ, or Ψ and Δ) is minimised by well known procedures like those of Newton or Levenberg-Marquardt [42]. Where d_{layer} and the substrate dielectric function is known, ε_{layer} can be determined in an analogous manner. In the general case, however, with both d_{layer} and ε_{layer} unknown, the determination of these parameters is much more difficult because only two parameters are measured for each photon energy. Guidelines in treating this problem are:

- ε_{layer} should be similar for layers of comparable thicknesses
- Im(ε_{layer}) below the fundamental gap should be small
- the structure of the substrate optical gaps should not remain in ε_{layer}

As regards the last requirement, the optical gaps are specific to the material and will, in general, vary in energy. If d_{layer} is too large or too small, ε_{layer} will include substrate features. This criterion is very stringent and it will not always be possible to eliminate the substrate contribution. For example, if the substrate dielectric function is inaccurate, or if an interlayer is formed during growth, attempts to eliminate substrate structures in ε_{layer} may not be successful.

3.3.2.3 Thin Layer Approximation

For very thin layers, the effective dielectric function as calculated in the three-phase model can be approximated by (following the notation of Sect. 1.3) [43]:

$$<\varepsilon> = \varepsilon_b + \frac{i4\pi d_{layer}\varepsilon_0^{1/2}}{\lambda} \frac{\varepsilon_b(\varepsilon_b - \varepsilon)(\varepsilon - \varepsilon_0)}{\varepsilon(\varepsilon_b - \varepsilon_0)} \left(\frac{\varepsilon_b}{\varepsilon_0} - \sin^2\phi_0\right)^{1/2} \quad (3.12)$$

$$\approx \varepsilon_b + \frac{i4\pi d_{layer}}{\lambda} \varepsilon_b^{3/2} \quad (3.13)$$

where ε is the dielectric function and d_{layer} the thickness of the overlayer, and ε_b and ε_0 are the dielectric functions of the substrate and ambient, respectively. This approximation holds for layer thicknesses up to approximately 0.6 nm. Equation 3.13 follows if $|\varepsilon_b| \gg |\varepsilon| \gg \varepsilon_0$, a condition which is often fulfilled with semiconductors as substrates, and transparent overlayers. Remarkably, ε has dropped out of equation completely, which means that films can be detected and removed without having to be identified. In the case of surfaces, however, the meaningful quantity is not the overlayer dielectric function ε but, instead, the product $\varepsilon \cdot d_{layer}$. Thin layer approximations are also discussed in Sect. 4.2.

3.3.3 Inhomogeneous Materials

For substrates or layers composed of different materials, (old) effective medium theories like those of Maxwell-Garnett [44], Looyenga [45] or Bruggemann [46] may be used. An effective dielectric function, $<\varepsilon>$, can be calculated from known dielectric functions of the constituents, and for special topologies (e.g.

Maxwell-Garnett: isolated spheres in a matrix). For these effective medium theories to be applicable, the dimensions for all constituents (except the matrix material) must be small compared to the wavelength of light, but large enough to have the same dielectric function as the bulk material. For rough surfaces, where the last requirement is not strictly fulfilled, some authors have successfully fitted their spectra with the Bruggemann effective medium theory [47].

In the simple situation of particles with dielectric function, ε_p, embedded with a volume fraction, f, in a matrix, m, with dielectric function, ε_m, these theories allow $<\varepsilon>$ to be calculated using:

Maxwell-Garnett
$$\frac{<\varepsilon>-\varepsilon_m}{<\varepsilon>+2\varepsilon_m} = f\frac{\varepsilon_p-\varepsilon_m}{\varepsilon_p+2\varepsilon_m} \qquad (3.14)$$

Looyenga
$$\sqrt[3]{<\varepsilon>} = f\sqrt[3]{\varepsilon_p}+(1-f)\sqrt[3]{\varepsilon_m} \qquad (3.15)$$

Bruggemann
$$f\frac{\varepsilon_p-<\varepsilon>}{\varepsilon_p+2<\varepsilon>} = (1-f)\frac{<\varepsilon>-\varepsilon_m}{\varepsilon_m+2<\varepsilon>} \qquad (3.16)$$

Whether the assumption that the particles (and the matrix) have the bulk dielectric function is reasonable depends on the material of the particles, and their size. Where diameters are so small that quantum size effects occur, the dielectric function may differ significantly from the corresponding bulk value.

As regards the topology, the Bergman theorem can be used to decide if the simple theories are applicable [48, 49]:

$$<\varepsilon> = \varepsilon_m\left(1-f\int_0^1 g(n)dn/(t-n)\right) \qquad (3.17)$$

where $t = \varepsilon_m/(\varepsilon_m-\varepsilon_p)$. The function $g(n)$ describes the topology of the system and is independent of the dielectric functions of the constituents. This formula has the advantage that the percolation can be varied independently of the volume fraction [43, 49]. If the wavelength of light is of the order of the dimensions of the constituents, the situation is much more complicated. The samples produce scattered light, which is usually unpolarised or partially polarised. There are no general theories available for this case. The other limiting case, where the wavelength is smaller than the dimensions of the particles, has been intensively investigated for the IR spectral region [50].

3.4 Selected Examples

In this section the capabilities of ellipsometry in the study of surface and interface properties are discussed. Examples of the change of the dielectric function with reconstruction, and by adsorption of gases or metals, are given. Ellipsometry is also sensitive to buried layers and the section ends with an example of such an *ex situ* study. The effect of sample temperature is considered first, as it is important in all these studies because the dielectric function is strongly temperature-dependent.

3.4.1 The Temperature Dependence of the Dielectric Function

The temperature dependence of ε allows the temperature of the sample surface to be determined by SE. Sample temperature measurement is a particular problem in growth chambers. In most cases thermocouples are used for this purpose, but are often placed at some distance from the sample surface, resulting in a temperature measurement deviating significantly from that of the sample. One reason for such a separation of sample surface and thermocouple is that a thermocouple could induce contamination. Temperature determination with SE has been performed, with accurate results [51].

Where a sample temperature is varied, for example in a growth process, changes of the dielectric function have to be considered. For the semiconductors with the diamond or zincblende structure, an increase in temperature induces a red shift in the spectral position of the optical gaps, due to a renormalisation of the states via the electron-phonon interaction [31].

Typical Si spectra for various temperatures are shown in Fig. 3.7. When the sample temperature is lowered, the optical gaps become much more sharper, the height of Im $<\varepsilon>$ near E_0', E_1 and E_2 increases, and the gaps shift to higher energy.

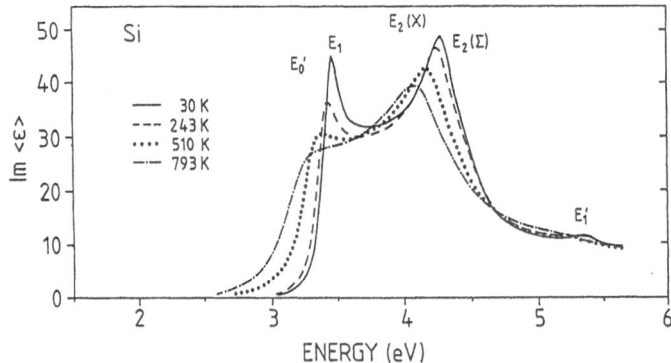

Fig. 3.7. Dielectric function of Si for various temperatures. Note that all features in the dielectric function are sharper at low temperatures [31].

3.4.2 Substrate Preparation

The measured dielectric function of a sample is very sensitive to substrate preparation. This is a problem where bulk dielectric functions must be determined. Conversely, this can be used to optimise substrate preparation. This is based on the fact that contamination and roughness lower the values of $Im<\varepsilon>$ near the optical gaps E_0' and E_2. The influence of an oxide film on $Im<\varepsilon>$ is demonstrated in a simulation shown in Fig. 3.8. The effective dielectric function of an oxidised sample is simulated using the dielectric functions of GaAs and its oxide [3, 52]. The height of $Im<\varepsilon>$ near the E_2-gap is particularly affected. This knowledge was used to find substrate preparations leading to microscopically abrupt surfaces with small amounts of contamination [3, 53, 54].

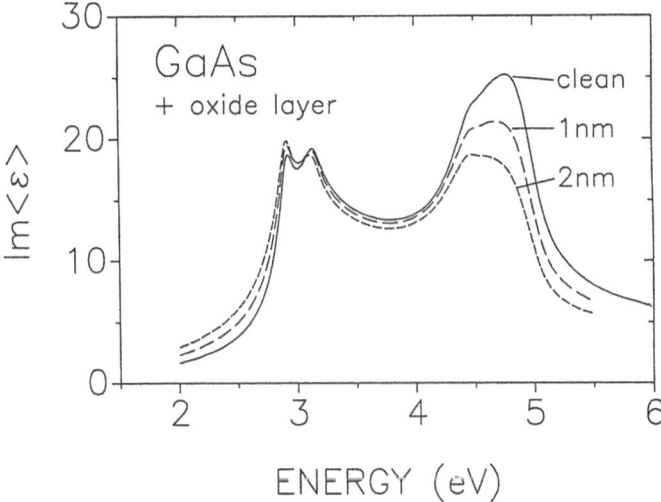

Fig. 3.8. $Im<\varepsilon>$ for GaAs, with oxide layers of thickness 1 nm (long dashes) and 2 nm (short dashes), calculated by a three-phase vacuum-oxide-bulk model, from the data of [3] and [52].

Furthermore, the substrate preparation can be followed *in situ*, as shown for Si in Fig. 3.9. This allows the etching procedure to be stopped when the best surface condition is reached.

It is also possible to identify the presence of surface films formed by certain preparation procedures. This was successfully done for the chemical treatment of HgCdTe and CdTe. In both cases, a thin film (below 1 nm) of amorphous Te is formed when using an HCl-based etch [56, 57]. Another application is the determination of the thickness of native oxide layers [58].

The creation of defects can also reduce the height of $Im<\varepsilon>$, especially near the gaps [3, 59, 60]. Defects were generated by ion bombardment methods, such as ion implantation [61] and plasma etching [62, 63]. With increasing defect density, $<\varepsilon>$ exhibits an increase in the broadening of the optical gaps and, finally, only a broad structure remains, typical of amorphous material [61, 64, 65].

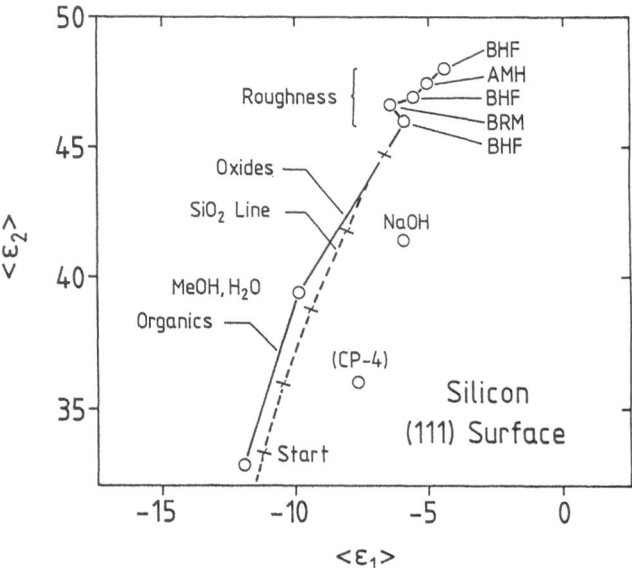

Fig. 3.9. Dependence of $<\varepsilon>$ on selected chemical reagents, for a natural film grown on a Syton-polished, (111)-oriented, Si crystal by a two year exposure to air. The measurement energy was 4.25 eV (after [55]).

Direct interpretation of $<\varepsilon>$ requires care, because contamination and oxidation of the surface may give incorrect results. Oxidation reduces the magnitude of the measured Im $<\varepsilon>$ value near E_2 and shifts the structures to lower energies, for most semiconductors (Fig. 3.8). Native oxide overlayers slightly influence the values for the energy and broadening of the optical gaps E_1 and $E_1 + \Delta_1$. The corresponding calculated values of $<\varepsilon>$ differ only slightly from those of clean bulk material. This is demonstrated by a simulation, using an effective dielectric function calculated for a 1.5 nm oxide layer on top of crystalline GaAs, following the procedure explained in Sect. 3.3.2. The values of $<\varepsilon>$ for GaAs and its oxide are taken from [3]. Fig. 3.10 shows the real part of $<\varepsilon>''$ for oxidised and clean (non-oxidised) GaAs.

The values obtained by a fit to the gaps are listed in Table 3.2. The calculated difference in the energy and broadening of $<\varepsilon>$, for clean and oxidised GaAs, can be seen to be small.

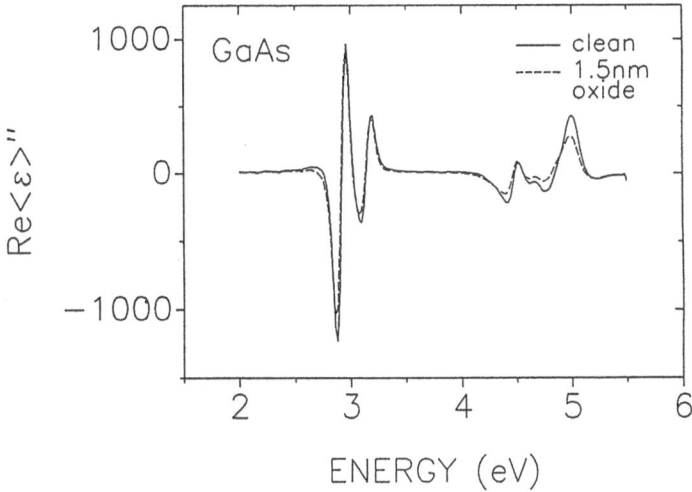

Fig. 3.10. Re $<\varepsilon>''$ for oxidised and clean (non-oxidised) GaAs. The energy and broadening of the gaps E_1 and $E_1 + \Delta_1$ are only slightly affected by the presence of an oxide overlayer.

Table 3.2. Energy and broadening of E_1 and $E_1 + \Delta_1$ for clean (non-oxidised) and oxidised GaAs, in eV (simulated).

	E_1		$E_1 + \Delta_1$	
	E_g (eV)	Γ (eV)	E_g (eV)	Γ (eV)
Clean	2.910	0.102	3.132	0.140
Oxidised	2.910	0.104	3.129	0.144

3.4.3 Surface Reconstruction

Many semiconductor surfaces are reconstructed. Due to a changed position of the atoms at and near the surface the electronics states near the surface differ from those of the bulk. Consequently, the dielectric function depends on surface reconstruction. The positions of the atoms near the surface deviate, however, only slightly with respect to that of the bulk. Therefore, the dielectric function of such a reconstructed surface is expected to be similar to the bulk dielectric function.

The surface dielectric function of thermally cleaned Si(001)2x1, and cleaved Si(111)2x1, has been determined (Fig. 3.11) [66]. Interestingly, both surface dielectric functions are similar to each other and resemble the dielectric function of bulk, crystalline Si. The sharp features of the E_0', E_1 and E_2-gaps, however, have vanished, and the absorption below E_0' has increased compared to the bulk [3]. In addition, an edge around 2 eV occurs in the surface dielectric function of Si(111)2x1 which might be caused by transitions involving surface states.

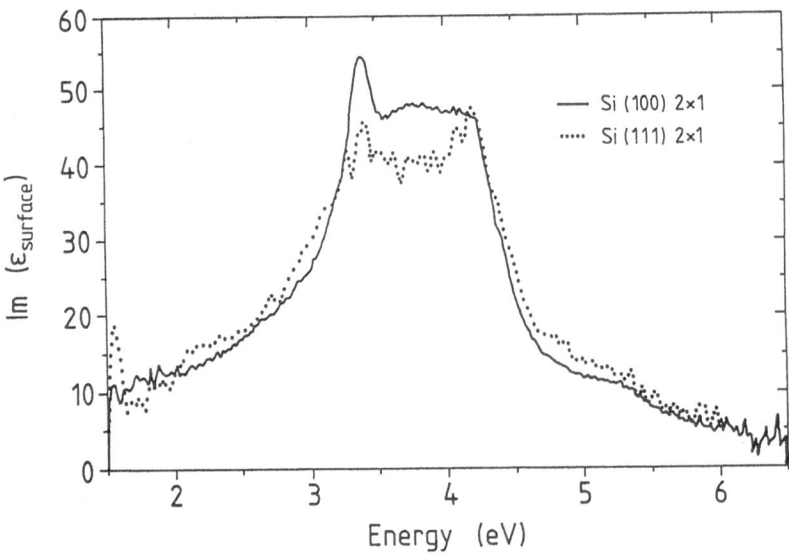

Fig. 3.11. Imaginary part of the surface dielectric function of Si(111)2x1 and Si(001)2x1. Note that both surface dielectric functions are similar [66].

For Ge(111), the irreversible reconstruction (2x1) → c(2x8) induced by annealing was studied by single wavelength ellipsometry (632.8 nm) [67]. A small change in the complex reflectance ratio was observed and correlated with the above transition. In a later publication [68], the reversible transition c(2x8)↔(1x1) was detected using SE. In Fig. 3.12, Re<ε> at a fixed wavelength of 600 nm close to the E_1-gap of bulk, crystalline Ge is shown as function of temperature. A change of the slope was noticed between 200°C and 250°C. This is the temperature range where the surface phase transition c(2x8)↔(1x1) was detected by low energy electron diffraction (LEED). For a polycrystalline, unreconstructed Ge layer, no change of the slope was observed. Therefore, the change of the slope of Re<ε> can be attributed to a change of the surface electronic properties due to a rearrangement of surface atoms during the c(2x8)↔(1x1) phase transition.

In Fig. 3.12, the intensity of the quarter-order LEED reflection peak, normalised to the low-temperature Debye-Waller factor [69], is compared with the SE data. The LEED intensity drops abruptly at around 300°C. On the other hand, the transition, as monitored by ellipsometry, spreads over a temperature range between 200°C and 300°C. This discrepancy might be explained by the fact that, if reconstruction occurs, surface and subsurface properties are changed and probed by ellipsometry, while LEED is most sensitive to the two topmost atomic layers.

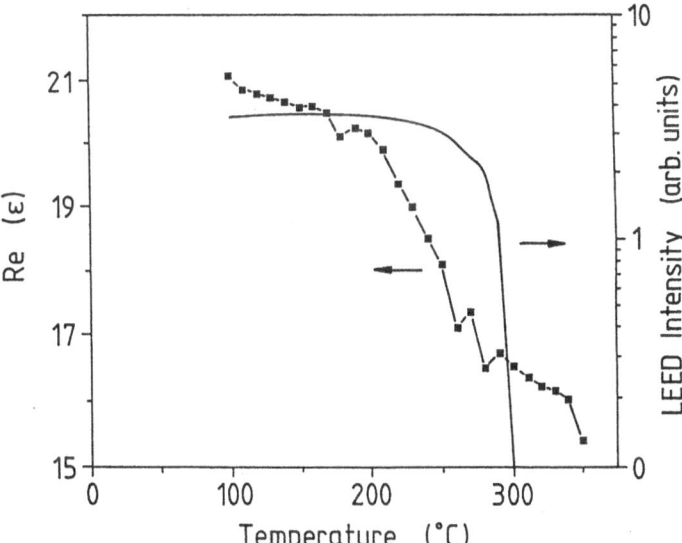

Fig. 3.12. Temperature behaviour of Re $<\varepsilon>$ (solid squares) and the intensity of a quarter-order LEED spot normalised to the low-temperature Debye-Waller factor [69].

From spectra taken before and after the transition, the surface dielectric function for Ge(111)2x8 and Ge(111)2x1 was derived using the standard three-phase (vacuum-surface layer-substrate) model [70]. One problem in the calculation is that the dielectric function depends on the sample temperature (Sect. 3.4.1). It is difficult to correct for this effect. In a first approximation, the whole spectrum was shifted in energy by an amount given by the shift of the dominant E_1-gap of Ge. Thus broadening of the gaps was neglected. The resulting surface dielectric functions are shown in Fig. 3.13. It can be concluded from the spectra that the main features of the surface density of states are not changed by the phase transition, in agreement with photoemission data [72, 73]. The rather strong variation of the ellipsometric signal, as shown in Fig. 3.12, could be explained by the relaxation of stress accumulated in the surface region, during heating, due to the different expansion coefficients of the bulk and the dense Ge(111)c(2x8)surface. These studies demonstrate the high sensitivity of the dielectric function and, consequently, SE to surface phenomena.

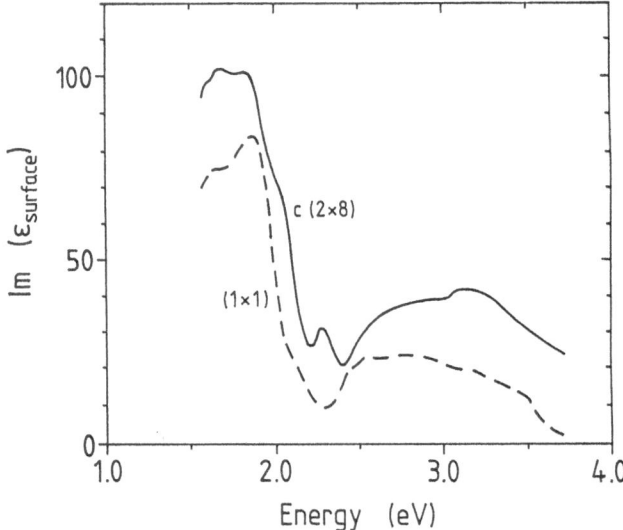

Fig. 3.13. Imaginary part of the surface dielectric function of the Ge(111) surface (dashed line) just above the c(2x8)↔(1x1) transition, and the corresponding function for the c(2x8) surface, calculated from data given by Zandvlieth *et al* [71] (solid line, values multiplied by 5). The thickness of the transition layer is assumed to be 0.5 nm (double layer of Ge).

3.4.4 Surface Change on Gas Adsorption

The adsorption of gas on surfaces has several important aspects. Firstly, the reconstruction of the surface might be changed by the adsorption of the gas. Secondly, it is possible to saturate danglind bonds at the surface or, more generally, to remove or create surface states. Measuring the dielectric function before and after saturation of the surface allows the contribution of the surface states to the dielectric function to be determined. This was performed for Si, Ge and GaAs, which were exposed to gases like Kr, O_2 and H_2 [74-76]. In addition, sticking coefficients of gases on surfaces may be determined [77]. As SE has submonolayer resolution, this can be done with high precision, as shown for several gases adsorbed on Si [77]. These early studies, however, were performed with single or multi-wavength ellipsometers.

The adsorption of hydrogen on Si has been, however, studied by SE. Hydrogen on Si(111) saturates the dangling bonds at the surface. This gives a nearly ideally-terminated Si surface with electronic states removed from the band gap position. Such a surface is thus very well suited for theoretical calculations and surface investigations. However, for hydrogen on Si(001), two phases exist. In the monohydride phase, the Si dimers are still present on the surface, while the single dangling bond per atom is saturated by hydrogen. The resulting LEED pattern is an overlay of (2x1) and (1x2) for double domain surfaces, or simply (2x1) for single domain surfaces, which can be produced with off-axis oriented Si(001) surfaces by annealing to high temperatures. In the dihydride phase, the Si dimer is replaced by hydrogen and thus two hydrogen atoms are bonded to each Si surface atom in the topmost layer. Hydrogen termination can be achieved either by adsorption of atomic hydrogen on a clean Si surfaces [66], or by wet chemical etching [78, 79]. The latter method works surprisingly well for (111) surfaces, leaving only small amounts of contamination [54]. For (001) surfaces, the results are somewhat poorer as the surface is left microscopically rough after etching.

The effect of hydrogen termination on the dielectric function of Si and Ge has been studied [66]. Atomic hydrogen was adsorbed on Si(001)2x1 (HF etched and heated to 900°C), and Si(111)2x1 (cleaved).

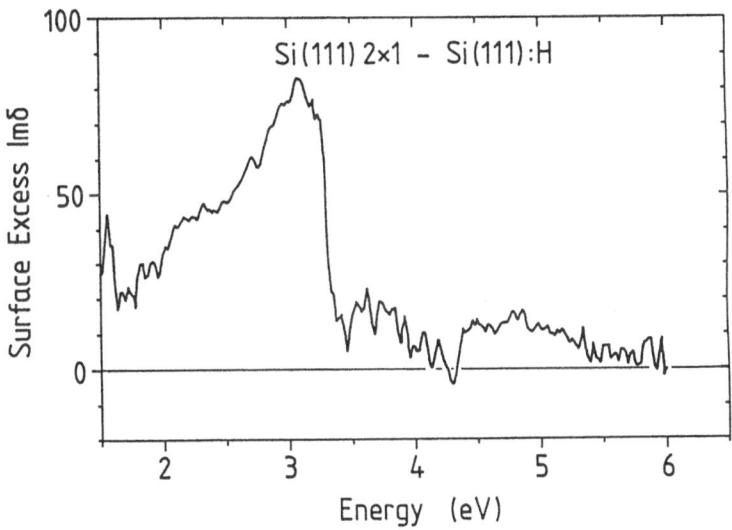

Fig. 3.14. Imaginary part of the surface excess function, determined from the SE spectra by comparing Si(111)2x1 with the hydrogen-terminated surface [66].

Figure 3.14 shows the effect of hydrogen termination on the optical response of Si(111)2x1. A broad maximum is observed near 3.1 eV and two small dips near 3.4 eV and 4.3 eV, close to the spectral positions of the E_1- and E_2-gaps of bulk Si. The imaginary part of the surface contribution drops to small values near the onset of bulk absorption (E_1-gap). No further features occur above 3.2 eV in the spectral region investigated. This indicates that possible contributions from surface states due to the hydrogen are not in this spectral region.

SE can, as shown, give some insight into the electronic structure at the surface. A real understanding of the results, however, can only be achieved by comparison of the data with theory.

3.4.5 Thin Metal Layers on Semiconductors

SE is very useful tool in the study of thin metal layers on semiconductors, from the submonolayer to the nanometer thickness range. This is of interest because metal contacts have to be made on all semiconductor devices, and metals also serve as dopants (Ga, In, Sb or As for Si). Furthermore, in the epitaxial growth of, for example, GaAs on a Si substrate, As bonds to Si and thus the first step in this heteroepitaxy is the formation of a metal monolayer. For contacts, these metal layers should be smooth and dense, and be without voids. Where metal atoms are used for doping, clustering of the metal atoms must be avoided. In consequence, the adsorption and clustering of metals on semiconductors is extensively studied.

3.4.5.1 Gallium Adsorption on Si(111)

Andrieu and d'Avitaya have analysed the adsorption of Ga on Si(111) by ellipsometry and RHEED [80, 81]. Ga is used as dopant in MBE growth of Si and knowledge of how it adsorbs on Si is required. Dopant incorporation anomalies have been reported for both Sb and Ga [80, 82, 83], which may be explained, for Ga, by nucleation during the adsorption. In this study, Ga adsorption was monitored in real time with a fixed wavelength. The samples were etched by the Shiraki procedure [84] and cleaned by heating to 700°C. Finally, a thin (10 nm) Si film was grown of this clean surface. All experiments started with a (7x7) reconstruction. For Ga growth using a cell temperature 700°C, resulting in a flux of about 10^{12} atoms $cm^{-2}s^{-1}$, the Si substrate temperature was varied from room temperature (RT) to 700°C.

For the substrate at RT, there is a linear decrease in $\cos \Delta$ and $\tan \Psi$ up to 0.7 ML, and above 0.7 ML with a different slope (Fig. 3.15) [85]. The RHEED pattern is (7x7) and changes to a diffuse (1x1) pattern when 0.7 ML Ga is deposited. This means that Ga is not adsorbed in the same way below and above 0.7 ML. First Ga adsorbs on Si sites and then, after 0.7 ML is adsorbed, only Ga adsorption sites are available.

Three regimes were found when the substrate temperature is between 400° and 500°C (Fig. 3.16). From 0 to 1/3 ML there is little effect on $\cos \Delta$ and $\tan \Psi$. The (7x7) pattern disappears at about 0.1 ML, where a ($\sqrt{3}x\sqrt{3}$) pattern takes its place. Between 1/3 and 2/3 ML $\cos \Delta$ and $\tan \Psi$ again vary linearly with adsorption and, by 2/3 ML Ga coverage, the pattern changes again to a diffuse (1x1) structure. Gallium desorbs after the shutter is closed, as indicated by the change in $\cos \Delta$ and $\tan \Psi$, as well as the reappearance of the ($\sqrt{3}x\sqrt{3}$)

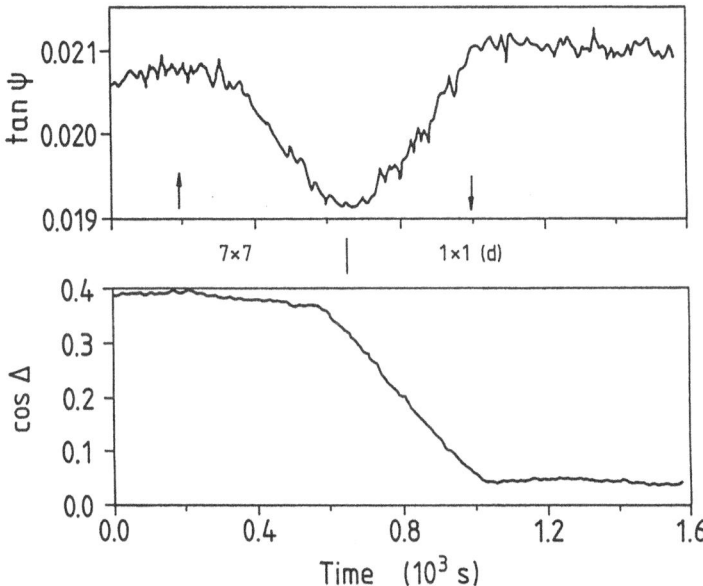

Fig. 3.15. The variation of $\cos\Delta$ and $\tan\Psi$, with time, during Ga adsorption at 300°C. The arrows indicate the shutter opening and closing. The RHEED patterns are also indicated [85].

reconstruction. All this can be explained by the formation of Ga droplets above 2/3 ML. These droplets desorb when the substrate temperature is above 400°C. Moreover, a Ga phase transition below 2/3 ML occurs when the substrate temperature is raised above 400°C. It is well established that all dangling bonds of Si are saturated at 1/3 ML in the ($\sqrt{3}\times\sqrt{3}$) configuration [86, 87]. Consequently, the Ga adsorption sites are different below and above 1/3 ML, which leads to a change in the dielectric function [81].

3.4.5.2 Arsenic Layers on Silicon

The interest in As layers is twofold. Firstly, As forms ordered and stable monolayers on Si surfaces [88-90]. Secondly, *thick* As overlayers can be used to protect surfaces against contamination for several weeks [91-95]. This method of protection has been used in the growth of II-VI compound semiconductors on III-V substrates. A III-V buffer layer is grown and capped by As. The As is then desorbed in a second growth chamber, where the II-VI compound is grown. Arsenic-capped surfaces can also be used in surface studies because, after desorption of As, a clean surface is revealed for further investigations. Eventually, small crystals of arsenic oxides are formed in the presence of oxygen, which leads to a destruction of the cap [95, 96].

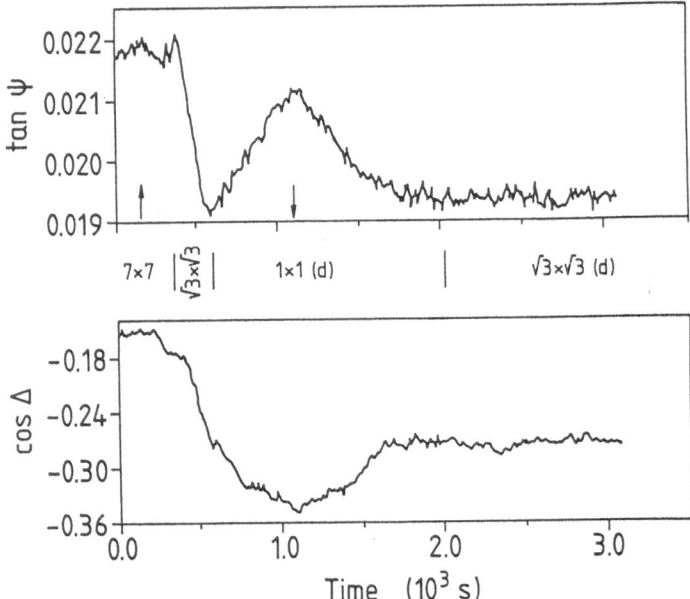

Fig. 3.16. The same experiment as in Fig. 3.15, but for adsorption at 500°C. The values of cos Δ and tan Ψ are little affected until a Ga coverage of 1/3 ML, where a linear decrease in both parameters occur. This ends when the shutter is closed (indicated by the downward arrow). The ($\sqrt{3}$x$\sqrt{3}$) pattern does not disappear [85].

Figure 3.17 shows a typical spectrum for a thick, freshly grown As layer on Si [97]. The broad structure around 2.5 eV indicates the amorphous structure of the As. The energetic position of the maximum and the shape of Im $<\varepsilon>$ compare well with that of amorphous As [94]. After desorbing the As cap at temperatures lower than 300°C, an ordered As monolayer is obtained [96]. This monolayer can be desorbed at temperatures around 700°C [88].

Arsenic evaporation and desorption has been monitored by ellipsometry at a fixed wavelength of 300 nm (4.13 eV), as shown in Fig. 3.18. This spectral position is close to the E_2 transition of bulk Si and provides the lowest penetration depth. A value of $<\varepsilon>$ was taken every 20 s, where the time resolution of the ellipsometer for typical values of the analyser rotation frequency, and the number of averaged cycles, is about 13 s.

Arsenic is deposited on a thermally cleaned Si surface from an MBE cell operating at about 300°C. The pressure in the chamber rises to 10^{-6} mbar, corresponding to an As flux of greater than 1 ML s^{-1}. At 1 ML, a "plateau" (A) in Im $<\varepsilon>$ is reached. More As then adsorbs, but no new plateau is reached. The sticking coefficient of As on As is very low, requiring a large flux to grow layers of As on Si at RT. Finally, an equilibrium coverage (B) is reached, where the number of desorbing As molecules equals the number of adsorbing molecules. The equilibrium coverage is strongly dependent on the sample temperature and As flux, since the desorption rate and the sticking coefficient are functions of temperature. In this case a fit yields a thickness of 1.3 nm of amorphous As [97].

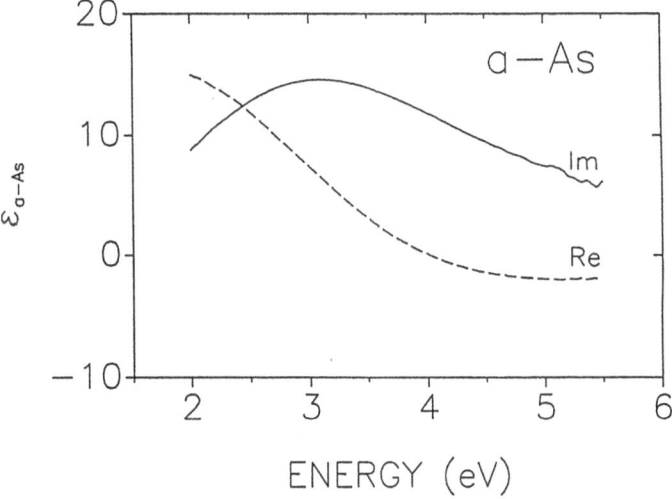

Fig. 3.17. The dielectric function of amorphous As obtained from an As-capped Si surface. The broad structure in Im $<\varepsilon>$ indicates the amorphous character of the layer [97].

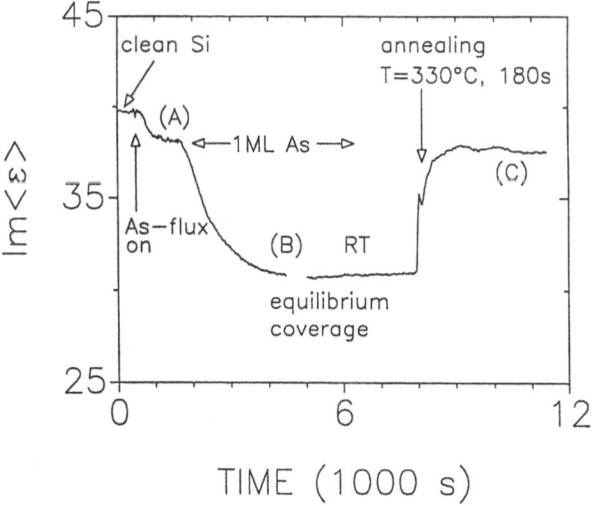

Fig. 3.18. Arsenic deposition on, and desorption from, Si(001). Arsenic was deposited on the Si surface at 33°C. At 1 ML As, a plateau (A) in Im $<\varepsilon>$ is reached. Further deposition produces a saturation thickness of a few ML (B). On stopping the As flux, no change in Im $<\varepsilon>$ occurs. Arsenic was then desorbed by raising the sample temperature to 330°C. On cooling the sample to room temperature (C), Im $<\varepsilon>$ returns to the plateau value (A), as expected [97].

The strong temperature dependence of the As layer thickness is also known from growing caps. At RT, the sticking coefficient of As is too low, and the samples have to be cooled below 0°C to achieve As layers of several nm thickness [98].

The transient in Im$<\varepsilon>$ during the desorption of the As layers is shown in Fig. 3.18. It is found that, after stopping the As evaporation and without raising the temperature, Im$<\varepsilon>$ does not change. Increasing the temperature leads to an immediate increase in Im$<\varepsilon>$, accompanied by an increase of the chamber pressure. This demonstrates that most of the As is loosely bound. Desorption ends when approximately 250°C is reached, in agreement with desorption of the As caps on Si. The new saturation level (C) in Im$<\varepsilon>$, which is reached when the sample cools down to RT, is nearly the same as the plateau in (A).

For quantitative analysis of the As desorption, the temperature dependence of $<\varepsilon>$ has to be considered (Sect. 3.4.1). The E_2-gap of Si shifts to lower energies, and broadens, as the sample temperature increases. Since the broadening lowers Im$<\varepsilon>$ more than the energy shift can compensate for, there should be a decrease in Im$<\varepsilon>$ with increasing temperature of the sample. The origin of the dip in Im$<\varepsilon>$ during the desorption of As is still unclear, and might be caused either by the desorption process or by a slight misalignment of the sample holder during annealing (due to differing thermal expansions). The adsorption and desorption of As were reproduced several times with similar results. The first plateau can consequently be identified as that state where 1 ML has adsorbed on the Si surface.

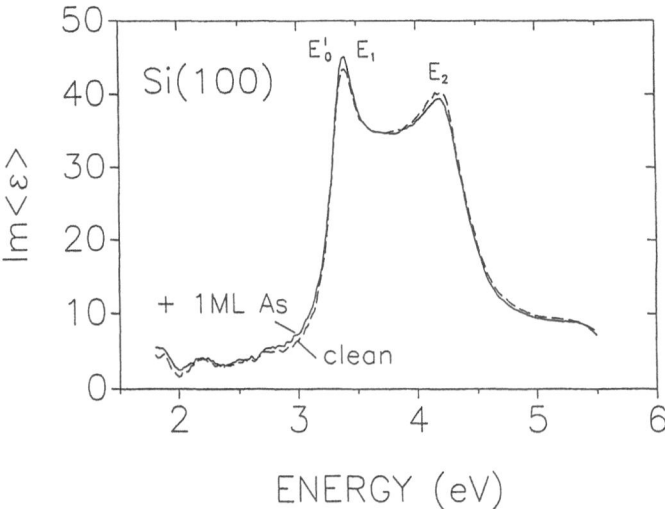

Fig. 3.19. Im $<\varepsilon>$ for 1 ML As on Si(001) (solid line) compared to that for the clean sample (dashed line).

The dielectric function of 1 ML of As on Si(001), and that of the corresponding clean surface, are shown in Fig. 3.19. The surface of this wafer is slightly misoriented (3° off, towards [011]). Therefore, the surface is microscopically rough and the height of Im $<\varepsilon>$ near the E_2-gap is reduced compared to literature data for bulk Si [3].

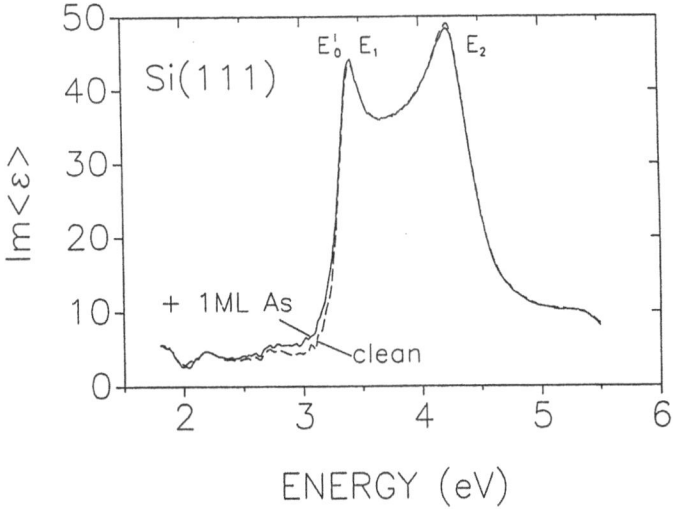

Fig. 3.20. Im $<\varepsilon>$ for 1 ML As on Si(111) (dashed line) compared to that for the clean sample (solid line) [97].

The effect of the As ML on the dielectric function is small but significant. The measured difference in Im $<\varepsilon>$ agrees well with calculations [99]. The corresponding dielectric function of an As ML on Si(111) is shown in Fig. 3.20. As with As on Si(001), the effect of the As ML is small, and greatest near the gap features.

The layer dielectric function, ε_{layer}, was calculated in order to separate the substrate influence (Fig. 3.21). A feature in Im(ε_{layer}) occurs near 4 eV. The spectral position of this feature is close to a peak in the polarisability in the z-direction (normal to the surface) calculated by Del Sole and co-workers [100] (see Sect. 2.3.2). This peak originates from transitions of the lone pair state of As to antibonding backbond states of As-Si.

That the small differences in Fig. 3.19 are indeed significant can easily be judged from the signal-to-noise (S/N) ratio in Fig. 3.18. It is clear from this figure that SE has submonolayer resolution. However, on-line monitoring of the desorption process would be possible only if the time resolution of the ellipsometer was significantly reduced.

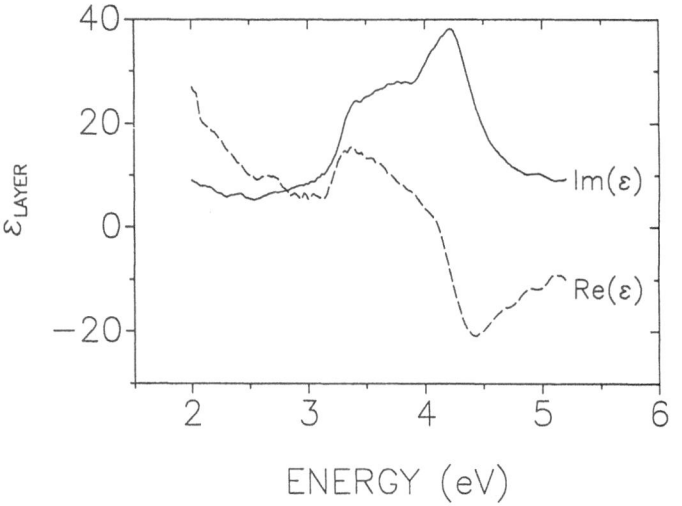

Fig. 3.21. Dielectric function of the As monolayer evaluated with the three-phase model from the data in Fig. 3.20, and a layer thickness of 0.3 nm [97]. The meaningful quantity for surfaces is the product of ε_{layer} and d_{layer}, but here only the lineshape is important.

3.4.5.3 Antimony Layers on Si and GaAs

The growth mode of Sb layers on Si and GaAs surfaces has been extensively studied by classical surface sensitive techniques like scanning tunneling microscopy (STM) [101, 102] or extended x-ray absorption fine structure (EXAFS) [103], and optical methods such as RS [104] or SE [105-107]. In contrast to many other metals, Sb is non-reactive and forms abrupt interfaces on both kinds of substrate. From RS experiments it is well known that the first ML of Sb grows epitaxially on GaAs(110) and, thus, results obtained for this ordered ML are well suited for comparison with theory. On evaporating further Sb, an amorphous layer grows, followed by an amorphous to crystalline phase transition when a critical layer thickness is reached [104, 108]. The barrier height at the interface depends on the structure and morphology of the layers [104], and investigations during the growth of Sb layers are especially interesting in the context of the unresolved Schottky .barrier formation problem.

In a detailed study [105, 106], Sb was deposited stepwise on cleaved GaAs(110). After each evaporation step a spectrum was taken and an overlayer dielectric function was calculated using the method described in Sect. 3.3.2. Figure 3.22 shows the dielectric functions obtained for the Sb layers with increasing thickness.

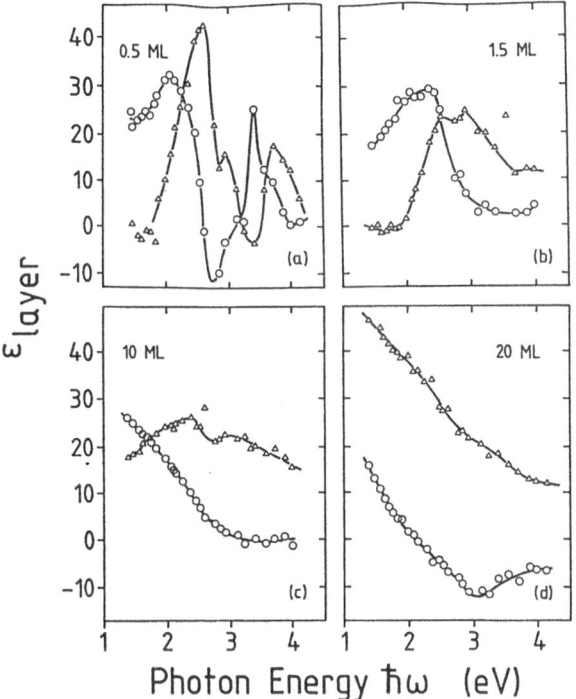

Fig. 3.22. Re(ε_{layer}) (circles) and Im(ε_{layer}) (triangles) of Sb overlayers on GaAs(110) vs. photon energy, for various Sb layer thicknesses: (a) 0.5ML; (b) 1.5 ML; (c) 10 ML; (d) 20 ML [106].

For submonolayer coverages, Im(ε_{layer}) exhibits maxima near 2.6, 3.0 and 3.7 eV. When 1.5 ML Sb are deposited, the shape of the dielectric function is dominated by a broad maximum located between 2.5 eV and 3.0 eV. This maximum disappears with increasing thickness until Im(ε_{layer}) reveals semimetallic character for a 20 ML deposit. Similarly, the prominent feature in Re(ε_{layer}) for 0.5ML disappears with increasing thickness.

The growth mode is slightly different for Sb on GaAs(001). No ordered ML is observed and an amorphous layer grows by an approximately layer-by-layer mode [109]. The Sb layer also crystallises at a critical thickness. A similar behaviour is observed when Sb is evaporated on Si(001), Si(110), and Si(111). The development of Im $<\varepsilon>$ with increasing layer thickness is shown in Fig. 3.23. For the highest coverage, the light hardly penetrates the Si substrate and the effective dielectric function is similar to that of bulk Sb. The layer thicknesses were obtained when calculating the layer dielectric function (Sect. 3.3.2.2).

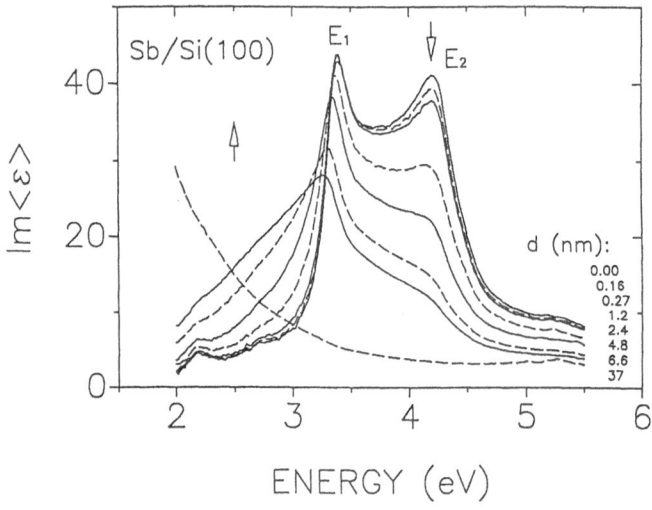

Fig. 3.23. The development of the effective dielectric function for Sb on Si(001) with increasing layer thickness [107].

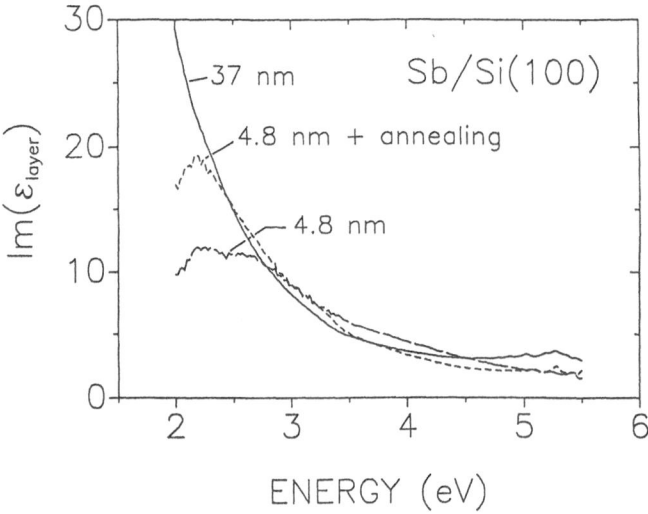

Fig. 3.24. Layer dielectric functions for two different coverages (4.8 and 37 nm) obtained from the data in Fig. 3.23. The dielectric function of the 4.8nm layer is also shown after annealing at 200°C for 20 min [107].

The layer dielectric function, ε_{layer}, at 4.8 and 37 nm, derived from the effective dielectric function, $<\varepsilon>$ (Fig. 3.23) is shown in Fig. 3.24. For comparison, $Im(\varepsilon_{layer})$ for the 4.8nm layer after annealing is added. Before annealing there is only a broad maximum in $Im(\varepsilon_{layer})$ indicating that this layer is still amorphous. After annealing, the shape of the layer dielectric function changes and becomes similar to that of the 37 nm layer. The absolute values still differ, most likely due to size effects in the Sb layers. Above 37 nm, size effects are unimportant as the absolute values of the dielectric function are close to those of bulk crystalline Sb.

The crystallisation of the Sb layer was also studied in real time as Sb was being evaporated continuously. The change of $Re<\varepsilon>$ and $Im<\varepsilon>$ with time was recorded at a fixed wavelength of 300 nm (4.13 eV) close to the E_2-gap of Si (Fig. 3.25). An edge in $Im(\varepsilon_{layer})$ is observed at around 15 nm, indicating the phase transition from amorphous to crystalline Sb. The thickness of the Sb-layer when the phase transition occurs corresponds well with values reported in the literature for different substrates [110]. Surprisingly, the crystallisation process is not instantaneous, but takes some minutes. This is consistent with crystallisation propagating laterally, or along the surface normal. RS results reveal that amorphous and crystalline Sb regions coexist.

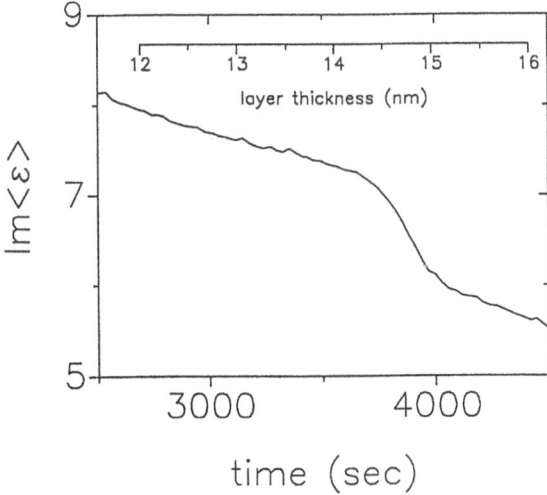

Fig. 3.25. $Im<\varepsilon>$ measured at 4.13 eV, as a function of time during Sb deposition. A step-like decrease in $Im<\varepsilon>$ is observed at a thickness of around 15 nm, corresponding to the phase transition to the metallic D_{3d} structure [107].

The growth of Sb layers can thus be divided in three steps. Firstly, in the submonolayer and ML régime, the dielectric function of the layers are determined by sharp structures that are probably related to the electronic structure of the interface, and can be interpreted as interband transitions. The ML is found to be well ordered for (110)-oriented GaAs substrates, but not for (001). Secondly, for layers below a critical thickness of around 15 nm, the optical properties are dominated by those of amorphous Sb. The layers do not exhibit the semimetallic character of Sb, or the sharp features of submonolayer films. In the third stage of growth after crystallisation, the optical properties of the layers are similar to that of polycrystalline (textured) Sb. However, size effects are found in layers thinner than 37 nm.

3.4.6 Heterostructure Interfaces

Knowledge of the properties of semiconductor heterostructures, and the optimisation of growth conditions, is essential in producing semiconductor devices like lasers. Numerous publications deal with these systems. Here, we show that SE is a powerful tool for characterising semiconductor heterostructures in general, and heterostructure interfaces in particular. A good description of how to extract layer properties for multi-quantum-well (MQW) structures or superlattices from SE has been published [111].

As an example, we consider a MQW-structure grown by metal-organic-vapour-phase-epitaxy (MOVPE), consisting of 50 periods of InGaAs wells and InP barriers. In addition, a 5 nm InP spacer was introduced after every 5th double layer. A 50 nm capping layer was grown on top. Growth interruptions were applied at all heterointerfaces to allow for smoothing by rearrangement of the atoms. Growth interruptions of 2s were also performed when switching from InGaAs to InP. During this time, InGaAs was stabilised by AsH_3 and then by PH_3. Thus the formation of quaternary InGaAsP alloy was likely to occur at this interface. The growth was interrupted for 1s at the inverted interface. The InP surface was stabilised by PH_3 during this time. Thus only an InGaP monolayer is expected at this interface, due to the different bond lengths and lack of chemical reaction.

The measured dielectric function for the MQW structure is shown in Fig. 3.26. Above the $E_1 + \Delta_1$ gap, the spectrum is similar to InP. Below the E_1 gap, however, a shoulder is visible at the low energy side with a spectral position above the $E_1 + \Delta_1$ gap of InGaAs. Its energy is close to that determined for InGaAsP. This indicates that an exchange reaction of the group V element at the interfaces between InGaAs and InP has occurred. A simulation of $Im<\varepsilon>$ demonstrates this more clearly. The lineshapes of the simulated spectra without interlayers are significantly different from the measured spectra (Fig. 3.26). In contrast, an effective dielectric function calculated under the assumption that nearly all material in the InP barriers and the InGaAs quantum wells has reacted to form InGaAsP (leaving only one monolayer of InGaAs) reproduces the shape of the measured dielectric function very well. This indicates that quaternary material is formed during growth consuming all, or nearly all, of the ternary

InGaAs, in agreement PL measurements. The dielectric function of $In_{0.72}Ga_{0.28}As_{0.60}P_{0.40}$ was used for the simulation, indicating that the quaternary material formed must have a similar stoichiometry.

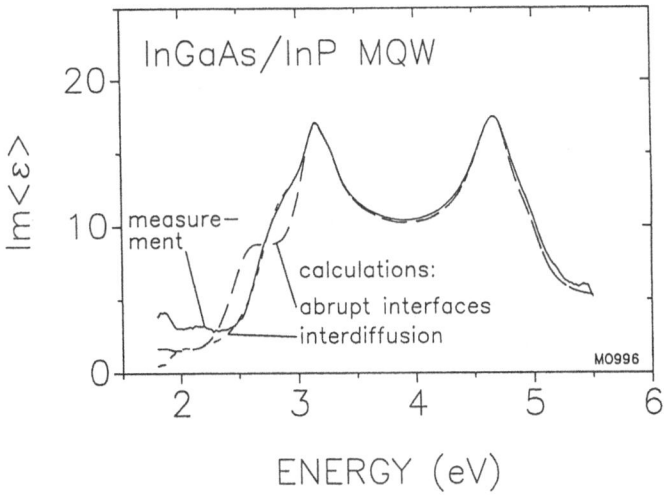

Fig. 3.26. Measured and simulated Im $<\varepsilon>$ for the MOVPE-grown MQW structure. The measured spectrum cannot be simulated without taking into account the formation of InGaAsP interlayers [112].

The interlayer formation is extremely strong for this sample, and was found to be weaker in other samples investigated. For a sample grown with a 50 nm spacer layer under similar conditions, the thicknesses of the interlayers were found (from the simulations of the dielectric function) to be 1 nm for the interfaces near to the surface and 4 nm for more deeply-lying interfaces. It is important to note that x-ray diffraction indicates that the two interfaces (InGaAs/InP and InP/InGaAs) are not the same. Interdiffusion appears to drive this interlayer formation in the MQWs investigated [112].

Interestingly, a quaternary interlayer of composition $In_{0.72}Ga_{0.28}As_{0.60}P_{0.40}$ would be lattice matched to InP (with an estimated error in composition determination of 10% at present). It thus appears likely that interdiffusion occurs in such a manner that the strain is minimised. The exchange of the group V element, alone, would give strained interlayers of InAsP and InGaAsP (both not lattice matched to InP) and is not in accordance with the experimental results. The Ga must diffuse into the InP after interface growth to obtain nearly lattice matched interlayers of quaternary material. This would reduce the strain caused by the exchange of the group V element, and is chemically favourable as the Ga-P bond is stronger than the Ga-As bond.

An interdiffusion process has also been observed for InAs/AlSb multilayer structures, which depended on how the interfaces were grown. When growing AlSb on an InAs layer, the sequence could be In-As-Al-Sb (AlAs-like) or As-In-Sb-Al (InSb-like) [113]. The second derivative of the effective dielectric function for such structures is shown in Fig. 3.27. The spectra are remarkable different for the two types of interfaces. Analysis of the spectra and comparison to RS data lead to the conclusion that interlayers were formed in the case of AlAs-like interfaces. In contrast, the InSb-like interfaces appeared to be stable.

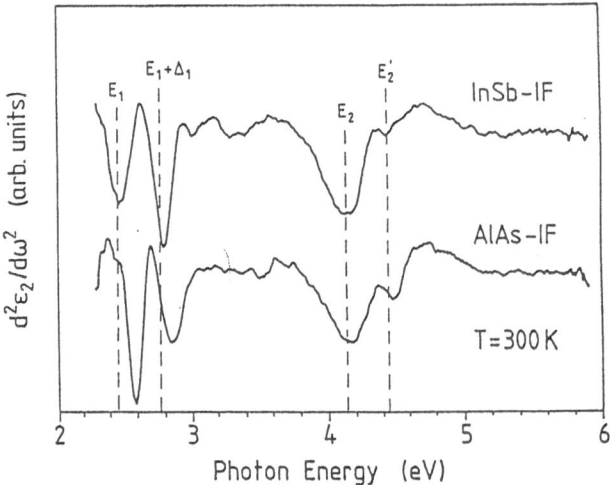

Fig. 3.27. Imaginary part of the second derivative of the effective dielectric function. The spectral positions of the gaps of the bulk materials are indicated [113].

References

1.   Jamin, J.: Annales de chimie et physique *29*, 263 (1850)
2.   Drude, P.: Annalen der Physikalische Chemie (Leipzig) *34*, 489 (1888)
3.   Aspnes, D.E., Studna, A.A.: Physical Review *B 27*, 985 (1983)
4.   Drevillon, B.: Thin Solid Films *163*, 157 (1988)
5.   Ferrieu, F., Dutatre, D.: Journal of Applied Physics *68*, 5810 (1990)
6.   Drevillon, B.: Applied Surface Science *63*, 27 (1993)
7.   Röseler, A.: Infrared Spectroscopic Ellipsometry. Akademie Verlag, Berlin 1990
8.   Azzam, R.M.A., Bashara, N.M.: Ellipsometry and Polarized Light. North Holland, Amsterdam 1977
9.   McCrackin, F.L., Passaglia, E., Stromberg, R.R., Steinberg, H.L.: Journal of Research of the National Bureau of Standards *67A*, 363 (1963)

10. Hauge, P.S.: Surface Science *96,* 108 (1980)
11. Aspnes, D.E.: Journal of the Optical Society of America *64*, 812 (1974)
12. Aspnes, D.E., Studna, A.A.: Applied Optics *14*, 220 (1975)
13. Hauge, P.S., Dill, F.H.: IBM Journal of Research and Development *17*, 472 (1973)
14. Jasperson, S.N., Schnatterly, S.E.: Review of Scientific Instruments *40*, 761 (1969)
15. Jasperson, S.N., Burge, D.K., O'Handley, R.C.: Surface Science *37*, 548 (1973)
16. see, for example: Super Quiet Xenon Lamps. Hamamatsu Technical Notes TN-106-03, August 1984
17. Hecht, E., Zajac, A.: Optics. Addison-Wesley, New York 1974, ch. 8
18. Aspnes, D.E.: Journal of the Optical Society of America *64*, 639 (1974)
19. Collins, R.W, An, I., Nguyen, H.V., Gu, T.: Thin Solid Films *206*, 374 (1991)
20. Kemp, J.C.: Polarized Light and its Interaction with Modulating Devices. Hinds International Incorporated, 5250 NE Elam Young Parkway, Hillsboro, OR 97124-6463 (1987)
21. Acher, O., Bigan, E., Drevillon, B.: Review of Scientific Instruments *60*, 65 (1989)
22. Azzam, R.M.A., Giardina, K.A., Lopez, A.G.: Optical Engineering *30*, 1583 (1991)
23. Boccara, A.C., Pickering, C., Rivory, J. (eds): Proceedings of the first conference on spectroscopic ellipsometry, Paris 1993. Thin Solid Films *233/234* (1993)
24. Johnson, R.L., Barth, J., Cardona, M., Fuchs, D., Bradshaw, A.M.: Review of Scientific Instruments *60*, 2209 (1989)
25. Barth, J., Johnson, R.L., Cardona, M., Fuchs, D., Bradshaw, A.M.: Physical Review *B 41*, 3291 (1990)
26. Aspnes, D.E., Quinn, W.E., Tamargo, M.C., Pudensi, M.A.A, Schwarz, S.A., Brasil, M.J.S.P., Nahory, R.E., Gregory, : Applied Physics Letters *60,* 1244 (192)
27. Calvani, R., Caponi, R., Cisternino, F.: Optics Communications *54* , 63 (1985)
28. Woollam, J.A., Snyder, P.G.: Journal of Applied Physics *62*, 4867 (1987)
29. Bassani, F., Pastori Parravicini, G.: Electronic States and Optical Transitions in Solids. Pergamon, Oxford 1975
30. Cardona, M.: Modulation Spectroscopy. Supplement 11 of Solid State Physics: Seitz, F., Turnbull, D., Ehrenreich, H. (eds.) Academic Press, New York 1969
31. Lautenschlager, P.: PhD dissertation. University of Stuttgart, Stuttgart 1987
32. Chelikowsky, J.R., Cohen, M.L.: Physical Review *B 14*, 556 (1976)
33. Aspnes, D.E. in: Balkanski, M. (ed.) Handbook on Semiconductors, vol. 2. North Holland, Amsterdam, 1980, p. 109
34. Burstein, F. in: Kubo, R., Kamimura, H. (eds.) Dynamical Processes in Solid State Optics. Syokabo Publishing Company and W.A. Benjamin Incorporated, New York 1967
35. Toyozawa, Y., Inoue, M.: Journal of the Physical Society of Japan *22*, 1337 (1967)
36. Rowe, J.E., Aspnes, D.E.: Physical Review Letters *25*, 162 (1970)
37. Aspnes, D.E., Kelso, S.M.: SPIE Proceedings *452*, 79 (1983)
38. Aspnes, D.E., Kelso, S.M., Logan, R.A., Bhat, R.: Journal of Applied Physics *60*, 754 (1986)
39. Yao, H., Woollam, J.A., Wang, P.J., Tejwani, M.J., Iterovitz,S.A.A.: Applied Surface Science *63,* 52 (1993)
40. Logothetidis, S., Polatoglou, H.M., Ves, S.: Solid State Communication *68*, 1075 (1988)
41. Rossow, U.: Journal of Luminescence *57*, 205 (1993)
42. Press, W.H., Flannery, B.P., Teukolsky, S.A., Vetterling, W.T.: Numerical Recipes in C: The Art of Scientific Computing. Cambridge University Press, Cambridge 1988, ch. 9 and 15
43. Apsnes, D.E.: Thin Solids Films *89*, 249 (1982)
44. Maxwell-Garnett, J.C.: Philosophical Transactions of the Royal Society of London *203* , 385 (1904). *Ibidem 205A,* 237 (1906)

45. Looyenga, H.: Physica *31*, 401 (1965)
46. Bruggemann, D.A.G.: Annalen der Physikalische Chemie (Leipzig) *24*, 636 (1935)
47. Aspnes, D.E.: SPIE Proceedings *452*, 60 (1983)
48. Bergman, D.: Physics Reports *C43*, 377 (1978). *Ibidem*: Bulk physical properties of composite media. Les methodes de l'homogeneisation, Edition Eyrolles (1985)
49. Apsnes, D.E.: Physical Review *B 33*, 677 (1986)
50. Theiss, W.: PhD dissertation. Rheinisch-Westfälisch Technische Hochschule, Aachen 1989
51. Cong, Y., Collins, R.W., Messier, R., Vedam, K., Epps, G.F., Windischmann, H.: Journal of Vacuum Science and Technology *A 9*, 1123 (1991)
52. Aspnes, D.E., Schwartz, B., Studna, A.A., Derick, L., Koszi, L.A.: Journal of Applied Physics *48*, 3510 (1977)
53. Palik, E.D., Bermudez, V.M., Glembocki, O.J.: Journal of the Electrochemical Society *132*, 871 (1985)
54. Lewerenz, H.J., Bitzer, T.: Journal of the Electrochemical Society *139*, L21 (1992)
55. Aspnes, D.E., Studna, A.A.: SPIE Proceedings *276*, 227 (1981)
56. Arwin , H., Aspnes, D.E.: Journal of Vacuum Science and Technology *A 2*, 1316 (1984).
57. Arwin , H., Aspnes, D.E.: Thin Solid Films *138*, 193 (1986)
58. Burkhard, H., Dinges, H.W., Kuphal, E.: Journal of Applied Physics*53*, 655 (1982)
59. Vedam, K., So, S.: Surface Science *29*, 379 (1972)
60. Vedam, K., McMarr, P.J., Narayan, J.: Applied Physics Letters *47*, 339 (1985)
61. Erman, M., Theeten, J.B., Chambon, P., Kelso, S.M., Aspnes, D.E.: Journal of Applied Physics *56*, 2664 (1984)
62. Rossow, U., Fieseler, T., Geurts, J., Zahn, D.R.T, Richter, W., Puttock, M.S., Hilton, K.P.: Journal of Physics: Condensed Matter *1*, SB231 (1989)
63. Rossow, U., Fieseler, T., Geurts, J., Zahn, D.R.T., Richter, W., Puttock, M.S., Hilton, K.P.: Journal of Physics Comndensed Matter *1*, SB231 (1989)
64. Rossow, U., Wagner, J., Richter, W.: unpublished work
65. Stuke, J., Zimmerer, G.: Physica Status Solidi *B 49*, 513 (1972)
66. Kelly, M.K, Zollner, S., Cardona, M.: Surface Science *285*, 282 (1993)
67. Quentel, G., Kern, R.: Surface Science *135*, 325 (1983)
68. Abraham, M., Le Lay, G., Hila, J.: Physical Review *B 41*, 9828 (1990)
69. Phaneuf, R.J., Webb, M.B.: Surface Science *164*, 167 (1985)
70. McIntyre, J.D.E., Aspnes, D.E., Surface Science *24*, 417 (1971)
71. Zandvlieth, H.J.W., van Silfhout, A.: Surface Science *195*, 138 (1988)
72. Aarts, J., Hoeven, A.J., Larsen, P.K.: Physical Review *B 38*, 3925 (1988)
73 Hricovini, K., LeLay, G., Abraham, M., Bonnet, J.: Physical Review *B 41*, 1258 (1990)
74. Dorn, R., Lüth, H: Physical Review Letters *33*, 1024 (1974)
75. Archer, R.J., Gobeli, G.W: Journal of the Physics and Chemistry of Solids *26*, 343 (1965)
76. Meyer, F., Kroes, A.: Surface Science *47*, 124 (1975)
77. Bootsma, G.A., Meyer, F.: Surface Science *14*, 52 (1969)
78. Chabal, Y.J.: Materials Research Society Symposium Proceedings *259*, 349 (1992)
79. Müller, A.B., Reinhardt, F., Resch, U., Richter, W., Rose, K.C., Rossow, U.: Thin Solid Films *233*, 19 (1993)
80. Andrieu, S., Arnaud d'Avitaya, F., Pfister, J.C.: Journal of Applied Physics *65*, 2681 (1989)
81. Andrieu, S., Arnaud d'Avitaya, F., Pfister, J.C.: Surface Science *238*, 53 (1990)
82. Iyer, S.S., Metzger, R.A., Allen, F.G.: Journal of Applied Physics*52*, 5608 (1981)
83. Kubiak, R.A., Newstead, S.M., Leong, W.Y., Houghton, R., Parker, E.H.C., Whall, T.E.: Applied Physics *A 42*, 197 (1987)

84. Ishizaka, A., Shiraki, Y.: Journal of the Electrochemical Society *133*, 666 (1986)
85. Andrieu, S., Arnaud d'Avitaya, F.: Journal of Crystal Growth *112*, 146 (1991)
86. Lander, J.J.Morrison, J.: Surface Science *2*, 553 (1964)
87. Nogami, J., Park, S., Poate, C.F.: Surface Science *203*, L631 (1988)
88. Uhrberg, R.I.G., Bringans, R.D., Olmstead, M.A., Bachrach, R.Z.: Physical Review *B 35*, 3945 (1987)
89. Olmstead, M.A., Bringans, R.D., Uhrberg, R.I.G., Bachrach, R.Z.: Physical Review *B 34*, 6041 (1986)
90. Patel, J.R., Golovchenko, J.A., Freeland, P.E., Gossmann, H.-J.: Physical Review *B 36*, 7715 (1987)
91. Kowalczyk, S.P.Miller, D.L.Waldrop, J.R.Newman, P.G.Grant, R.W.: Journal of Vacuum Science and Technology *19*, 255 (1981)
92. Etienne, P., Alnot, P., Rochette, J.F., Massies, J.: Journal of Vacuum Science and Technology *B 4*, 1301 (1986)
93. Bernstein, R.W., Borg, A., Husby, H., Fimland, B.-O., Grande, A.P., Grepstadt, J.K.: Applied Surface Science *56-58*, 74 (1992)
94. Schäfer, B.J., Förster, A., Londschien, M., Tulke, A., Werner, K., Kamp, M., Heinecke, H., Weyers, M., Lüth, H., Balk, P.: Surface Science *204*, 485 (1988)
95. Resch, U., Esser, N., Raptis, I., Waßerfall, J., Förster, A., Westwood, D.I., Richter, W.: Surface Science *269/270*, 797 (1992)
96. Wilhelm, H., Richter, W., Rossow, U., Zahn, D.R.T., Woolf, D.A., Westwood, D.I., Williams, R.H.: Surface Science *251/252*, 556 (1991)
97. Rossow, U., Frotscher, U., Richter, W., Zahn, D.R.T.: Surface Science *287/288*, 718 (1993)
98. Woolf, D.A., University of Wales College of Cardiff; Förster, A., KFA Forschungszentrum Jülich, private communications
99. Shen, T.H., Matthai, C.C.: Surface Science *287/288*, 672 (1993)
100. Reining, L., Del Sole, R., Cini, M., Ping, J.G.: Physical Review *B 50*, 8411 (1994)
101. Martensson, P., Meyer, G., Amer, N., Kaxiras, E.M., Pandey, K.C.: Physical Review *B 42*, 7230 (1990)
102. Elswijk, H.B., Dijkkamp, D., van Loenen, E.J.: Physical Review *B 44*, 3802 (1991)
103. Woicik, J.C., Kendelewicz, T., Miyano, K.E., Cowan, P.L., Bouldin, C.E., Karlin, B.A., Pianetta, P., Spicer, W.E.: Physical Review *B 44*, 3475 (1991)
104. Pletschen, W., Esser, N., Münder, H., Zahn, D., Geurts, J., Richter, W.: Surface Science *178*, 140 (1986)
105. Strümpler, R., Lüth, H.: Surface Science *182*, 545 (1987)
106. Strümpler, R., Lüth, H.: Thin Solid Films *177*, 287(1989)
107. Rossow, U., Frotscher, U., Esser, N., Resch, U., Müller, Th., Richter, W., Woolf, D.A., Williams, R.H., Applied Surface Science *63*, 35 (1993)
108. Hünermann, M., Geurts, J., Richter, W.: Physical Review Letters *66*, 620 (1991)
109. Resch-Esser, U., Frotscher, U., Esser, N., Rossow, U., Richter, W.: Surface Science *307-309*, 597 (1994)
110. Fuchs, G., Melinon, P., Santos Aires, F., Treilleux, M., Cabaud, B., Hoareau, A.: Physical Review *44*, 3926 (1991)
111. Erman, M., Alibert, C., Theeten, J.B., Frijlink, P., Catte, B.: Journal of Applied Physics *63*, 465 (1988)
112. Rossow, U., Krost, A., Werninghaus, T., Schattke, K., Richter, W., Hase, A., Künzel, H., Roehle, H.: Thin Solid Films *233*, 180 (1993)
113. Spitzer, J., Fuchs, H.D., Etchegoin, P., Ilg, M., Cardona, M., Brar, B., Kroemer, H.: Applied Physics Letters *62*, 2274 (1993)

Acknowledgements

It is a pleasure to thank U. Frotscher and P. Marsiske for drawing the figures. We are glad to present unpublished results from a collaboration with J.Wagner (FHI Freiburg). Special thanks are due to D. E. Aspnes for delivering his data.

Chapter 4.
Reflection Difference Techniques

Dietrich Zahn

Technische Universität Chemnitz-Zwickau, D-09107 Chemnitz

4.1 Introduction

Reflectance measurements are conventionally used to derive the optical constants of solid materials, i.e. the refractive index, n, and the extinction coefficient, κ. At near normal incidence, for example, the reflectance, R, (the ratio of the reflected to incident EM intensity: see Sect. 1.3) is given by:

$$R = \frac{(n-1)^2 + \kappa^2}{(n+1)^2 + \kappa^2} \qquad (4.1)$$

The optical constants are related to the dielectric function, $\varepsilon = \varepsilon' + i\varepsilon''$, by:

$$\varepsilon' = n^2 - \kappa^2 \qquad \text{and} \qquad \varepsilon'' = 2n\kappa \qquad (4.2)$$

Both real and imaginary part of the dielectric function can be obtained by Kramers-Kronig analysis of the reflectance data. The features of the dielectric function in the spectral range of interest from the near IR (~ 1 eV) to the near UV (~ 6 eV) are usually induced by interband optical transitions. Consequently, reflectance measurements may be employed for the identification of critical points in the electronic bandstructure of the material (for this purpose, however, SE is often a more appropriate technique, as discussed in Sect. 3.3.1). The information depth of the reflectance measurement is determined by the penetration depth, $d_p = \lambda/4\pi\kappa$, (where λ is the wavelength) of the EM radiation into the solid. For a typical semiconductor such as GaAs, the penetration depth, d_p, is plotted in Fig. 4.1 for photon energies from 2 to 6 eV. It is apparent that d_p always exceeds 5 nm and is, in fact, much larger in the visible. As a result, conventional reflectance measurements provide information which almost entirely originates from the bulk of the material. The surface contribution to the reflected intensity can reach percentage levels only in the near UV region.

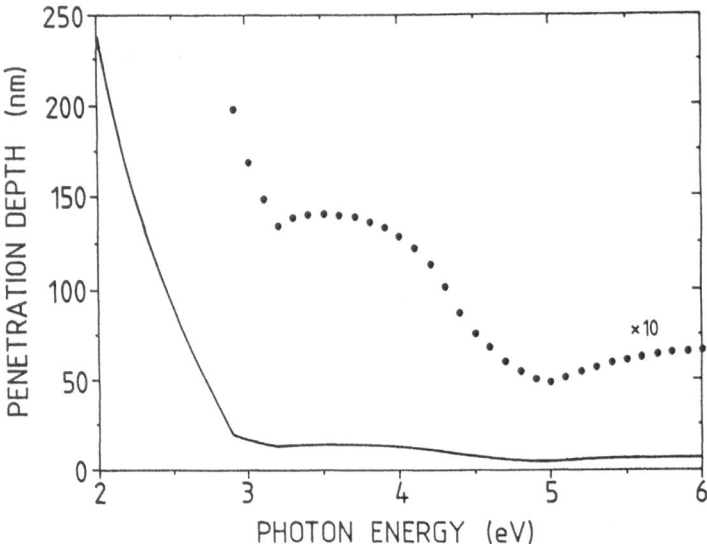

Fig. 4.1. Penetration depth of light, d_p, as a function of photon energy (data derived from [3]).

The bulk sensitivity prevents ordinary reflectance measurements from detecting and monitoring surface-related phenomena which occur, for instance, during the epitaxial growth of semiconductors. In this case, when the modification of the surface dielectric properties is of interest, the normally dominating bulk contribution has to be suppressed. The sensitivity then needed should be of the order of 1 ML (~0.25-0.5 nm), or better. This required surface sensitivity can be achieved by utilising difference techniques.

The influence of a thin surface film, whose thickness, d, is very much less than the EM wavelength ($d/\lambda \ll 1$: for 1 ML, $d/\lambda \sim 10^{-3}$, typically), on the reflectance from a substrate was discussed by McIntyre and Aspnes [1]. The fractional change in the p-polarised reflectance at near-normal incidence may be defined as:

$$\frac{\Delta R_{pp}}{R_{pp}} = \frac{R_{pp}(d) - R_{pp}(0)}{R_{pp}(0)} \tag{4.3}$$

where $R_{pp}(d)$ and $R_{pp}(0)$ are the reflectance from the substrate with and without the surface film, respectively. Using the *isotropic* three-phase model ($\varepsilon_{xx}^{film} = \varepsilon_{yy}^{film} = \varepsilon^{film}$ in Sect. 1.3), we have, assuming an ambient with $\varepsilon_0 = 1$:

$$\frac{\Delta R_{pp}}{R_{pp}} = -\frac{8\pi d}{\lambda} \cdot \text{Im}\left[(1 - \varepsilon^{film})/(1 - \varepsilon_b)\right] \tag{4.4}$$

where ε^{film} and ε_b are the dielectric functions of the surface film and the (isotropic) bulk, respectively. Model calculations for monolayer films on metal and dielectric substrates are given in Ref. 1. The Authors demonstrate that the sensitivity can be further enhanced by detecting the reflectance changes at oblique angles near the Brewster angle.

4.2 Surface Differential Reflectance

One of the first experimental approaches to studying the surface properties of semiconductors by differential techniques was presented by Chiarotti, Chiaradia, and co-workers. An overview of these experiments was given by Selci *et al* [2]. They measured the difference in reflectance between a clean semiconductor surface (R_{pp}^{clean}), usually cleaved in UHV, and an oxidised surface (R_{pp}^{ox}) at near-normal incidence. The structure of the clean semiconductor surface usually differs significantly from the truncated bulk, leading to the presence of electronic surface states which again affect the reflectance of the sample. The aim of such experiments is to obtain the resulting surface contribution to the dielectric function. Thus the clean sample is also treated in such a way as to incorporate a surface layer, distinct from the bulk and accounting, *inter alia*, for the effect of surface states. When the sample is oxidised and has an oxide layer with thickness d_{ox}, the oxidation usually leads to a removal of the surface states as can be monitored, for instance, by valence band photoemission spectroscopy.

The Authors measured an SDR signal defined as (see equation 1.13):

$$\left.\frac{\Delta R}{R}\right|_{SDR}^{0°} = \frac{R_{pp}^{clean} - R_{pp}^{ox}}{R_{pp}^{ox}} \tag{4.5}$$

which can be approximated by

$$\left.\frac{\Delta R}{R}\right|_{SDR}^{0°} \approx \frac{R_{pp}^{clean} - R_{pp}^{sub}}{R_{pp}^{sub}} - \frac{R_{pp}^{ox} - R_{pp}^{sub}}{R_{pp}^{sub}} \tag{4.6}$$

where R_{pp}^{sub} is the reflectance of the truncated bulk, with $d \rightarrow 0$, as mentioned above. However, this quantity, R_{pp}^{sub}, is not accessible to the experiment. Using equation 4.6 in this form, on the other hand, allows both contributions to be expressed in terms of equation 4.4, which finally leads to:

$$\left.\frac{\Delta R}{R}\right|_{SDR}^{0°} = -\frac{8\pi d}{\lambda} \cdot \frac{(1-\varepsilon_b')(\varepsilon'' - \varepsilon_{ox}'') + \varepsilon_b''(\varepsilon' - \varepsilon_{ox}')}{(1-\varepsilon_b')^2 + (\varepsilon_b'')^2} \qquad (4.7)$$

where ε, ε_{ox}, and ε_b are the surface, oxide, and bulk dielectric functions, respectively. This equation is derived assuming that the layer thickness, d, is of the order of 1 nm and is approximately equal for the oxide and the surface layer. The case of different thicknesses can be taken into account by introducing a minor modification to equation 4.7. Very often, oxide layers are transparent and thus lack the presence of electronic transitions in the energy range of interest (from about 1 to 6 eV): in this case $\varepsilon_{ox}'' \approx 0$ is a valid assumption. Equation 4.7 can then be written in the simple form:

$$\left.\frac{\Delta R}{R}\right|_{SDR}^{0°} = d \cdot \left[A\varepsilon'' - B(\varepsilon' - \varepsilon_{ox}')\right] \qquad (4.8)$$

where A and B contain only bulk properties

$$A = \frac{8\pi}{\lambda} \cdot \frac{(\varepsilon_b' - 1)}{(1-\varepsilon_b')^2 + (\varepsilon_b'')^2} \quad \text{and} \quad B = \frac{8\pi}{\lambda} \cdot \frac{\varepsilon_b''}{(1-\varepsilon_b')^2 + (\varepsilon_b'')^2} \qquad (4.9)$$

The quantities A and B can be calculated if the bulk dielectric function is known. Selci *et al* have done this using the data from Aspnes and Studna [3]. The result for Si, Ge, GaAs, and GaP is plotted in Fig. 4.2 for photon energies in the range from 0 to 4 eV. As can be seen from this figure, $B \approx 0$ in a large portion of the energy range and only differs from zero significantly near the higher energy limit. The reason for this is that the bulk semiconductors are hardly absorbing ($\varepsilon_b'' \approx 0$) below the direct band gap energies. Consequently, $\Delta R/R$ is directly proportional to $\varepsilon'' \cdot d$, and the real part of the surface contribution to the dielectric function can be obtained using the Kramers-Kronig relation.

The apparatus employed for SDR measurements is described in detail in Ref. 2. It allows an accuracy in $\Delta R/R$ of typically 1×10^{-4} to be achieved. As a representative example of an SDR measurement, Fig. 4.3 shows $\Delta R/R$ for a cleaved GaAs(110) sample after two different O_2 exposures. After a low dose of 1×10^4 Langmuir (L), $\Delta R/R$ exhibits an oscillatory behaviour which is

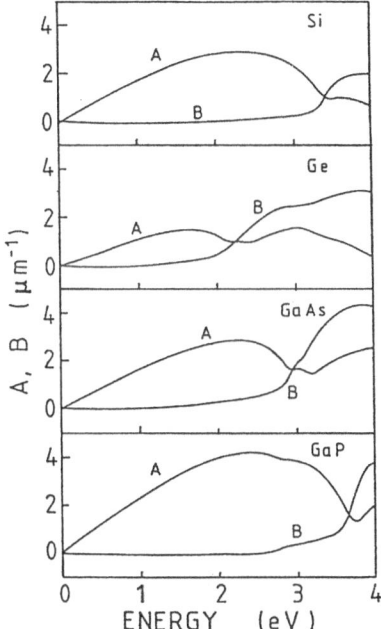

Fig. 4.2. Calculated energy dependence of A and B, used in equation 4.8, for Si, Ge, GaAs, and GaP using data from [3] (after [2]).

explained in terms of the Franz-Keldysh effect arising from O_2-induced band bending in the bulk of GaAs. This exposure, which corresponds to about 0.01 ML, has negligible effect on the surface states. They are, on the other hand, removed at the higher exposure of 1×10^6 L, where the SDR spectrum reveals a pronounced peak with the Franz-Keldysh oscillations superimposed. These oscillations, however, can be subtracted using the $\Delta R/R$ spectrum at the lower exposure in order to assess the surface induced change in reflectivity. The resulting SDR spectrum can then be converted into the surface dielectric function using equations 4.8 and 4.9 (for details, see Ref. 2). The imaginary part of the surface dielectric function derived in this way exhibits a peak around 2.9 eV, which is in qualitative agreement with theoretical predictions of transitions between dangling bond bands [4].

The evaluation of the dielectric function in this case is fairly complicated since $B > 0$ where the structure in the $\Delta R/R$ spectrum occurs. Thus $\Delta R/R$ is influenced by both real and imaginary parts of the surface dielectric function. Furthermore, the dielectric function of the oxides is also needed for the analysis. Nevertheless, this method is a very suitable tool for studying surface states at clean semiconductor surfaces, in particular when transitions at the surface involve

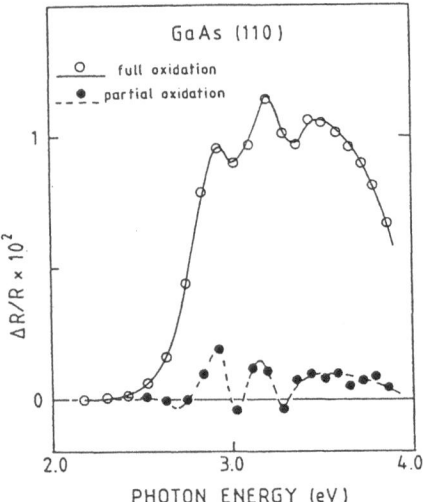

Fig. 4.3. $\Delta R/R$ versus photon energy for cleaved GaAs(110). The solid line and open circles refer to a complete oxidation (2×10^6 L; 1 ML coverage); dashed line and full circles refer to partial oxidation (1×10^4L) (after [2]).

energies below the bulk band gap, since the analysis is straightforward in this case.

In their SDR studies, Chiarotti, Chiaradia and co-workers have also demonstrated that the SDR signal is sometimes strongly anisotropic, i.e. dependent on the polarisation of the incident light. An impressive example is the below-band-gap anisotropy observed for Si(111)2x1 [5]. There $\Delta R/R$ shows a strong peak near 0.45 eV for light polarised along the [0$\bar{1}$1] direction of the sample while the signal disappears for polarisation along [$\bar{2}$11]. The data were explained using the anisotropic chain model proposed by Pandey for the Si(111)2x1 reconstruction [6]. Anisotropies, though weaker, were also detected for GaAs(110) and GaP(110) [7], as well as InP(110) [8] surfaces. These observations have probably stimulated the development of another technique, namely RAS, which will be discussed in the next section.

4.3 Reflection Anisotropy Spectroscopy

RAS was developed and improved by Aspnes and his co-workers to a standard which allows it to be used as a non-destructive, *in situ* optical monitor of growth processes [9]. It is applicable in UHV-based techniques like MBE or MOMBE, as well as in gaseous ambient techniques like MOVPE. The groups of Razeghi [10], and Richter [11], and Samuelson [12] also contributed an extensive effort to the rapid development of this technique within recent years.

4.3.1 Experiment

Considering a (001) oriented sample (the most commonly used growth surface), the near-normal-incidence reflection is measured for light polarised along the [110] and [$\bar{1}$10] directions. Since the bulk is isotropic for cubic semiconductor material, the reflection coefficient, r (Sect. 1.3), for the bulk is, in general, the same for the [110] and [$\bar{1}$10] directions: $r_{110} = r_{\bar{1}10}$. Thus, they cancel by subtraction and only contributions from the lower symmetry surface survive. The RAS signal may be written as:

$$\left.\frac{\Delta r}{r}\right|_{RAS} = 2\frac{r_{110} - r_{\bar{1}10}}{r_{110} + r_{\bar{1}10}} \tag{4.10}$$

or, if the reflectance, $R = |r|^2$ is measured, then

$$\left.\frac{\Delta R}{R}\right|_{RAS} = 2\frac{R_{110} - R_{\bar{1}10}}{R_{110} + R_{\bar{1}10}} \tag{4.11}$$

A possible realisation of the experiment, which allows the real as well as the imaginary part of the RAS signal to be evaluated, is shown in Fig. 4.4. This

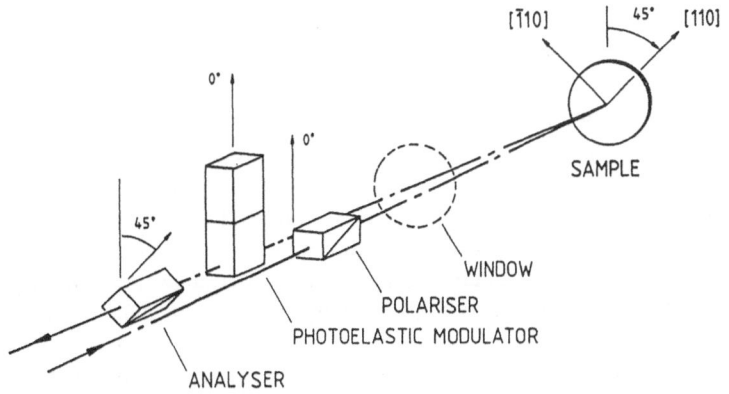

Fig. 4.4. Schematic diagram of an RAS set-up utilising a PEM (after [13]).

general configuration was also developed by the Aspnes group [13] and is now in use in several research laboratories around the world. A detailed discussion, together with a comparison of this configuration with those using rotating samples or analysers, is given in Ref. 13.

In detail, the equipment for this PEM configuration consists of a 75 W Xe short-arc lamp with an accessible photon energy range from 1.5 to 6 eV, front surface focusing optics, quartz, calcite or magnesium fluoride Rochon prisms, a PEM modulator operating at 50 kHz, a 0.1 m focusing grating monochromator, and an extended S20 photomultiplier. The RAS signal is then extracted using a phase sensitive lock-in amplifier. A typical RA spectrum can be recorded in 3 min [14], thus real-time monitoring of growth is hardly feasible utilising the full spectral range. The detection of changes in the RA signal at a fixed photon energy is much faster and can be performed within 100 ms. Consequently, RA spectra over the entire photon energy range can be employed to identify the photon energies for which maximum changes occur upon growth. These photon energies are then used for real time *in situ* monitoring of growth at fixed photon energies.

As can be seen from the alignment in Fig. 4.4, this system works as a near-null optical bridge. The light coming from the source and passing through the polariser can be divided up in equal parts as light polarised along [110] and [$\bar{1}$10]. If, in the absence of any anisotropy, $r_{110} = r_{\bar{1}10}$, then the linear polarisation of the incident beam is simply restored in the reflected beam, i.e. linearly polarised along the principal axes of the modulator. Consequently, the output is not modulated and the RA signal is zero.

A proper calculation of the output, however, yields a rather complicated expression, since misalignment of the optical components, as well as strain in the window, have to be included [13]. Fortunately, the real and imaginary part of the RA signal have different frequency dependencies and can therefore easily be distinguished by the lock-in detection. The voltage output reads:

$$\Delta V/V = 2[-\text{Im}(\Delta r/r) + \delta_1 \cos 2\gamma_1 + \delta_2 \cos 2\gamma_2 - 2a_p] \cdot J_1(\delta_C) \cdot \sin \omega t$$
$$+ 2[\text{Re}(\Delta r/r) + 2\Delta P + 2\Delta C] \cdot J_2(\delta_C) \cdot \cos 2\omega t \quad (4.12)$$

where ω is the modulation frequency, γ_1, γ_2, $(\Delta P - 45°)$, and $(\Delta C + 45°)$ are the reference azimuthal angles (see Fig. 4.4) of the incident beam window strain, the reflected beam window strain, the polariser, and the compensator (modulator) respectively, δ_1 and δ_2 and δ_C are the retardations due to the incident-beam window strain, reflected beam window strain, and the modulator, respectively, and J_1 and J_2 are Bessel functions. Two sets of parameters are needed to describe the window strain because it is usually not uniform over the window area. The effects of prism defects, including optical activity, are taken into account to first order by representing the transmitted mode for the polariser as $(x + ia_p y)$ [13]. The polariser and modulator azimuthal misalignments give rise to an offset in the real part, whilst the window birefringence and polariser imperfections affect the imaginary part of the RA signal. This may prevent absolute measurements. The sensitivity obtained with such a system was found to be about 2.5×10^{-5} in $Re(\Delta r/r)$, but about an order of magnitude poorer in $Im(\Delta r/r)$. This is essentially due to a coupling of residual mechanical vibrations to $Im(\Delta r/r)$ via non-uniform window strain. This problem can largely be removed by using a proper window design with very weak residual strain as, for instance, demonstrated for a UHV-MBE chamber [15].

The technique has been further improved by the Aspnes group by introducing an additional modulation, namely a slow rotation of the sample of approximately 0.1 Hz [16]. This approach allows systematic errors in the determination of the RA signal to be eliminated. Accurate values of $\Delta r/r$ may then be converted into the surface dielectric anisotropy (SDA), $\Delta \varepsilon \cdot d$, (with d being the thickness of the surface layer) using:

$$\Delta \varepsilon \cdot d = (\varepsilon_{110} - \varepsilon_{\bar{1}10}) \cdot d = i(\lambda/4\pi) \cdot (\varepsilon_b - 1) \cdot \Delta r/r \qquad (4.13)$$

where ε_b is the dielectric function of the bulk material and λ is the wavelength of the EM radiation. Precise data for ε_b are essential for the conversion of $\Delta r/r$ into SDA spectra.

4.3.2 Representative Examples

It is appropriate to begin this section with a look at those examples of growth studies which were performed in a UHV environment, since the RA spectra can be correlated with RHEED results. Considering the GaAs(001) surface, for instance, different reconstructions can be observed by RHEED depending on the Ga and As fluxes, and the substrate temperature [17]. At a typical growth temperature of 580°C one usually obtains (2x4) or (4x2) reconstructions if the surface is terminated by the As or Ga species, respectively. For the As-terminated surface, As-As dimers are aligned with the $[\bar{1}10]$ direction, while Ga-Ga dimers form along the [110] direction when the surface is Ga-stabilised. The alignment of such dimers should give rise to an anisotropy detectable by the RAS technique. This supposition was verified by Aspnes *et al* [18]. Figure 4.5 shows their data for the RA changes from As-terminated GaAs(001) to Ga- and Al-terminated surfaces. Considering first the transition As-to-Ga stabilised surface, a broad maximum can be seen at 2 to 2.5 eV. The authors assigned this change at this stage to the optical absorption merely induced by the presence of Ga-Ga dimers on the surface.

It was also demonstrated in this paper that the relative surface concentration of Ga-Ga dimers can be monitored by tracing the RA signal at a fixed wavelength in the region of the maximum anisotropy change of Fig. 4.5 (here 2.48 eV) as a function of time. This is shown in Fig. 4.6 where the Ga-stabilised surface was obtained by switching off the As flux for different lengths of time. For convenience the corresponding RHEED transients which were recorded

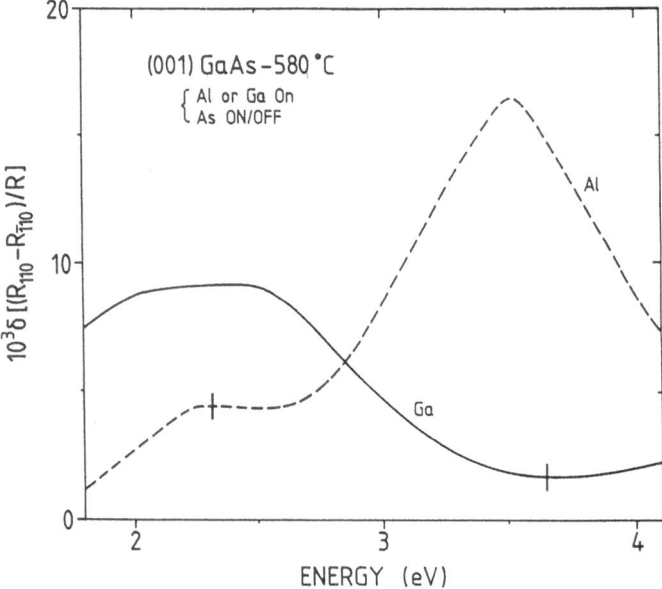

Fig. 4.5. Spectral dependence of changes in the RA signal, $\delta(\Delta R/R)$, when going from an As-terminated surface to either a Ga- or Al-stabilised surface (after [13]).

simultaneously are included. The overall resemblance of RA and RHEED transients is apparent. The RHEED intensity responds to surface structure while the RA intensity in this case reflects the relative concentration of As-As and Ga-Ga dimers on the surface. These distinct origins of the intensity changes explain the differences observed and illuminate the complementary nature of both techniques.

Considering again Fig. 4.5, the change from As-terminated GaAs(001) surface to a Al-terminated surface results in an entirely different RA response. Now a maximum is clearly visible around 3.5 eV, which is attributed to Al-Al dimers following the same argument as made above. As a result the RA measurement also enables to distinguish between the chemical species forming the surface dimers. Conventional RHEED, which provides merely structural information, lacks this capability.

When the growth of GaAs is monitored using a photon energy of around 2.5 eV the information obtained is related to the concentration of Ga-Ga dimers at the surface, i.e. to surface chemistry. If, on the other hand, a photon energy is chosen which lies outside the range of strong Ga-Ga dimer induced changes of the RA signal, for instance 3.5 eV, then RHEED and RA intensity transients are virtually identical [13]. In this case, both methods yield the same information on surface structure. RAS thus provides the unique potential of following both surface chemical and structural changes by monitoring the transients at different photon energies.

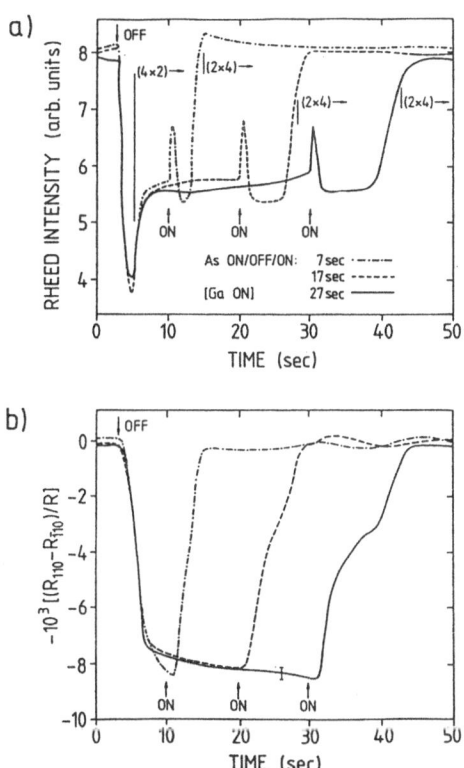

Fig. 4.6. (a) RHEED intensities for GaAs(001) for three different As-to-Ga-to-As surface stabilisation sequences, (b) simultaneously measured $\Delta R/R$ signal at 2.48 eV photon energy. The sample temperature was 580°C. The Ga flux remained on continuously at 1 ML/4.8 s. The As flux was interrupted as indicated. Changes in the RHEED pattern from the Ga-stabilised (4x2) to the As-stabilised (2x4) are also indicated by the vertical lines (after [18]).

Aspnes *et al* also recorded transients for the ternary compound $Al_{0.5}Ga_{0.5}As$ [13]. The evolution of the RA signal is then composed of GaAs- and AlAs-like contributions. A qualitative understanding of the adsorption mechanism for Ga and Al is obtained from these data. The sensitivity of such RA transients is approximately 0.1 s in time and 0.1 ML in surface coverage.

One of the most interesting features of RHEED is the observation of oscillations during growth. The period of these oscillations corresponds to the formation of one (001) atomic bilayer. Similar oscillations were also observed in the RA transients [19]. An example is shown in Fig. 4.7. Here the growth of AlAs on AlAs(001) was monitored using a photon energy of 3.44 eV, which is close to the strong Al-Al dimer induced change of the RA signal in Fig. 4.5. Therefore, the RA signal is sensitive to the surface chemistry. It was demonstrated that the RA oscillation period also correlates with the growth of exactly one bilayer. There is, however, a phase shift between the RHEED and RA oscillations. RHEED data are well known to exhibit phase delays related to the growth conditions, e.g. the As overpressure [20]. The RA oscillation, on the other hand, is not delayed in phase and due to the chemical sensitivity the completion of a cycle indicates that a monolayer of AlAs was deposited. The RA oscillations are very weak compared to the RHEED oscillations and signal averaging was employed in order to improve the S/N ratio. The weakness of the RA oscillations was explained by the As-stabilised growth conditions used, which probably led to an essentially As-stabilised surface over the entire growth cycle. Similar weak oscillations were also found for the growth of GaAs on GaAs(001), using a photon energy sensitive to surface structure [19].

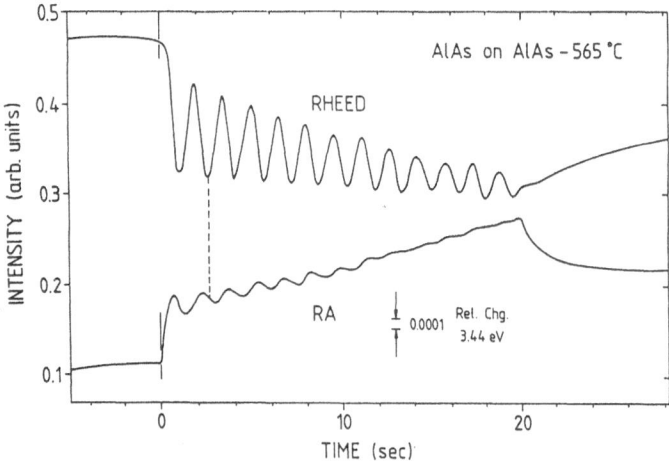

Fig. 4.7. Averages of nine RHEED and RA traces upon initiation of AlAs growth at 1.5 s per Al monolayer on an As-stabilised (2x4) AlAs surface (after [19]).

More recently, Aspnes *et al* presented data for the surface dielectric anisotropy (SDA) (see equation 4.13) for various reconstructions on GaAs(001) [16]. The variation of the imaginary part of the SDA from the (2x4) to the (4x2) reconstructions is shown in Fig. 4.8. Comparing this figure with Fig. 4.5, it is evident that the broad feature from 2.0 to 2.5 eV in the previous figure is now clearly resolved into two components, namely a positive peak at 1.8 eV and a negative one at 2.6 eV. The former is associated with the Ga-saturated (4x2) surface, while the latter is associated to the As-stabilised (2x4) surface. When going from the As- to the Ga-stabilised surface, the absorption along [$\bar{1}$10] decreases due to the decreasing As coverage, whilst the absorption along [110] increases because of the increasing Ga coverage. Their contributions both appear positive in the RA change from the As- to the Ga-stabilised surface because of the 90° orientation difference between the As-As and Ga-Ga dimers. Good agreement is found with tight binding calculations which attribute the peaks at 1.8 and 2.6 eV as being due to transitions between bonding Ga dimer orbitals and empty Ga lone pair states, and between filled As lone pair states and unoccupied As-As dimer antibonding orbitals, respectively [16].

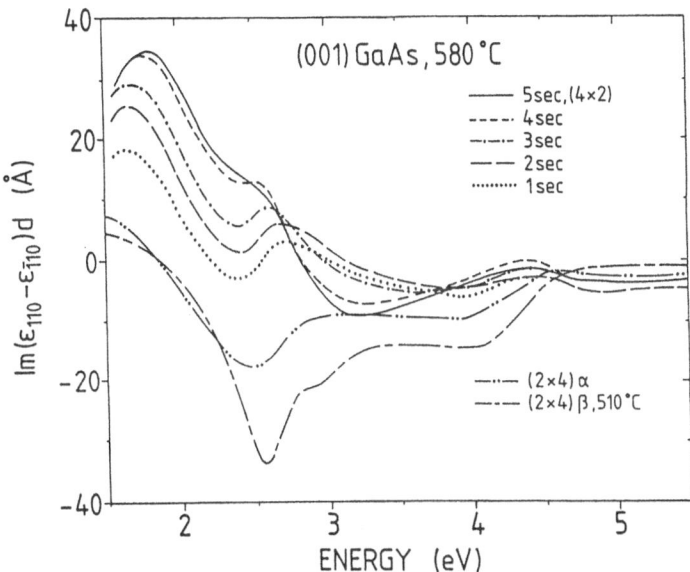

Fig. 4.8. Surface dielectric anisotropy spectra $\Delta(\mathrm{Im}\,\varepsilon \cdot d)$ of various reconstructions on GaAs(001) (after [16]).

Furthermore, *in situ* RAS is also very useful for monitoring the growth of strained layers, as demonstrated, for instance, for the MBE growth of InAs on GaAs(001) by Scholz *et al* [21]. These two semiconductors have a lattice mismatch of approximately 7% leading to a very small critical thickness of only a few monolayers. They interrupted growth at various stages and took RA spectra covering a photon energy range from 1.5 to 4.5 eV (limited by the UHV viewport transmission). Typical spectra of the real part of $\Delta r/r$ are shown in Fig. 4.9. The growth was initiated on a GaAs(001)-c(4x4) surface, as indicated by the

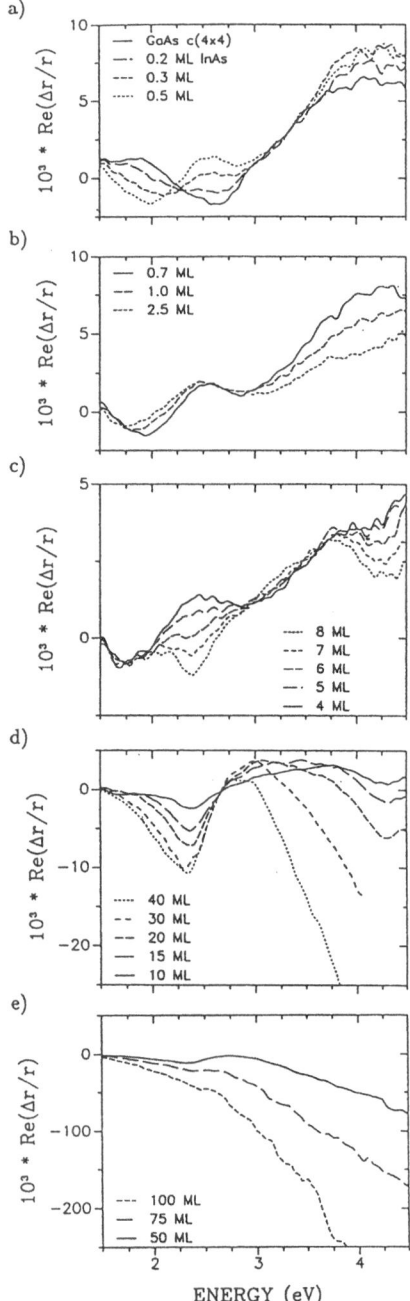

Fig. 4.9. RA spectra for InAs growth on GaAs(001) for different layer thicknesses of InAs increasing from a) to e) as noted in the figure. The spectra were taken while interrupting the growth for 2 minutes (after [21]).

simultaneously recorded RHEED pattern. At 0.5 ML of InAs, the reconstruction has changed to a (1x3) which results in pronounced changes in the RA spectra at around 2 and 2.5 eV (Fig. 4.9a). Intermediate InAs coverages lead to RA spectra which are linear combinations of those distinguished by the two distinct reconstructions. At coverages exceeding 0.5 ML (Fig. 4.9b), the changes in the RA spectra occur at the higher energy near 4 eV. In this regime, the RHEED pattern is transformed to (1x1) at $1\frac{2}{3}$ ML, which becomes spotty for higher coverages. This is believed to indicate the onset of three-dimensional growth after the critical thickness is exceeded. With further increase in coverage, a dip develops at 2.25 eV while an even stronger increase in anisotropy can be seen at high energies. This evolution of the RA spectra beyond the critical thickness was explained in terms of anisotropic surface roughness which increases with InAs thickness. This in particular affects the shorter wavelength part of the spectrum because of enhanced elastic scattering of light which also leads to an attenuation of the average reflectivity with increasing thickness.

Scholz *et al* used a three-phase model, as shown in Fig. 4.10, for simulating the RA spectra. In this model, the topmost layer is a rough anisotropic InAs layer of thickness, d_{l_1}, with a dielectric function, ε_{l_1}. Underneath is a dense layer of InAs with thickness d_{l_2} followed by a thick GaAs substrate. The dielectric functions

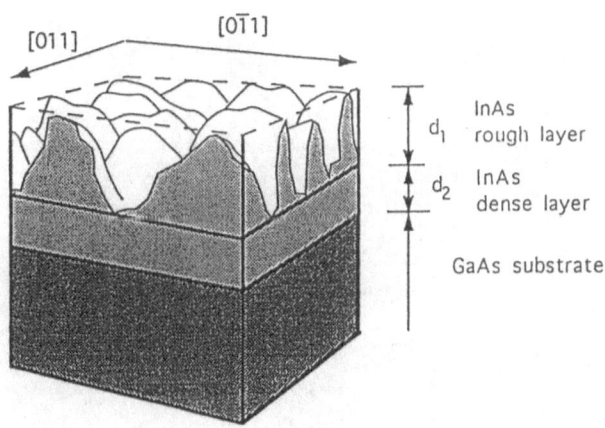

Fig. 4.10. Three-layer model used for simulating the RA spectra (after [21]).

ε_{sub} and ε_{l_2} of the substrate and the dense InAs layer, respectively, are those of the bulk materials, while ε_{l_1} is an effective dielectric function assuming that the effective medium consists of a fraction of dense crystalline InAs with dielectric function ε_{l_2} and a fraction f_a of the ambient vacuum with $\varepsilon_a = 1$, so that $f_{l_2} + f_a = 1$. According to effective medium theory [22], ε_{l_1} can be written as

$$\varepsilon_{l_1} = \frac{[q\varepsilon_a\varepsilon_{l_2} + (1-q)\varepsilon_h(f_a\varepsilon_a + f_{l_2}\varepsilon_{l_2})]}{(1-q)\varepsilon_h + q(f_a\varepsilon_{l_2} + f_{l_2}\varepsilon_a)} \qquad (4.14)$$

where ε_h is the dielectric function of the dominating material, i.e. either $\varepsilon_h = \varepsilon_{l_2}$ or $\varepsilon_h = \varepsilon_a$, using Maxwell-Garnett theory (Sect. 3.3.3).

The screening of the electric field through the rough layer is described by the depolarisation factor, q [23]. This factor is related to the surface geometry. In particular, roughness is likely to be anisotropic in the [110] and [$\bar{1}$10] directions of a (001) surface, as can be seen in scanning electron microscopy pictures, and this results in anisotropic depolarisation factors. Consequently, the Authors assumed that the depolarisation vectors are different for light polarised along the [110] and [$\bar{1}$10] directions. The dielectric function of the rough layer is then anisotropic since, from equation 4.14, it follows that:

$$\varepsilon_{l_1,110} = \varepsilon_{l_1}(\varepsilon_{l_2}, f_{l_2}, q_{110}) \qquad \text{and} \qquad \varepsilon_{l_1,\bar{1}10} = \varepsilon_{l_1}(\varepsilon_{l_2}, f_{l_2}, q_{\bar{1}10}) \qquad (4.15)$$

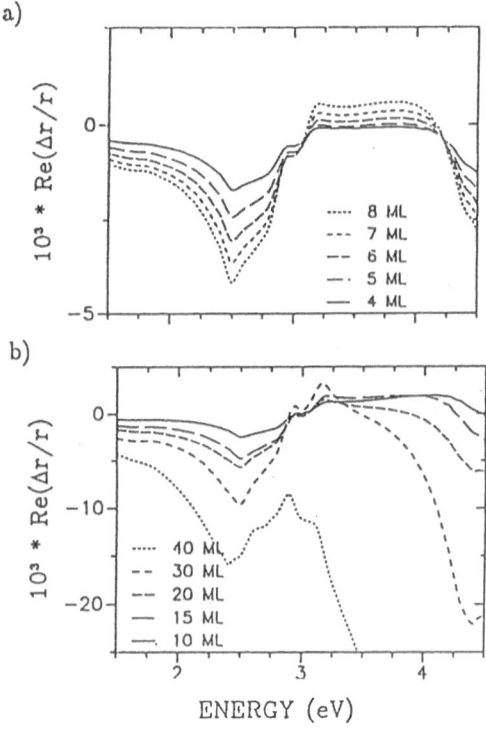

Fig. 4.11. Simulated RA spectra for InAs on GaAs(001) (after [21]).

Using this approach, the RA spectra were calculated for thicknesses exceeding 4 ML. The result is shown in Fig. 4.11. There is a clear resemblance with the corresponding experimental spectra of Fig. 4.9. It was thus concluded that the three-phase model delivers an appropriate description for the three-dimensional Stransky-Krastanov growth mode of InAs on GaAs(001).

A very similar optical model was successfully applied by Acher *et al* [10] for the analysis of RA spectra recorded during the MOVPE growth of lattice-mismatched semiconductors. The MOVPE growth technique excludes the application of electron-based growth analysis, like RHEED, because of the gaseous ambient. Furthermore, the situation on the surface is *a priori* more complex than for MBE. When growing, for example, GaAs by MOVPE, the group III element is supplied by metal-organic compounds such as trimethylgallium (TMG) or triethylgallium (TEG), while AsH_4 is usually used as the As precursor. Consequently, the metal-organic compound itself, or radicals thereof such as dimethylgallium or monomethylgallium and, of course, Ga itself, may be present on the growth surface. The pioneering work in RAS of MOVPE growth of GaAs was again carried out by Aspnes and his co-workers [24, 25].

The attachment of an RAS apparatus to a MOVPE reactor is straightforward since the reactors are usually built of quartz tubes. The influence of the quartz, which is not of optical quality, on the RAS signal is negligible. However, in conventional MOVPE when the precursors and their carrier gas (H_2 or N_2) are simultaneously introduced into the reactor, the walls are rapidly covered with reaction products and light attenuation causes a problem. Furthermore, the reactions may then also take place in the gas phase above the growing surface. Both problems can be avoided by using an atomic layer epitaxial (ALE) growth approach, where the precursors are supplied sequentially at low temperatures (370-450°C). Gas-phase reactions between the precursors are then eliminated, and the TMG molecule also stays intact in the gas phase ambient at these temperatures. Surface reactions involving the decomposition of TMG take place and are responsible for the growth of GaAs.

Despite the more complex situation in MOVPE, it was recently shown that RA spectra of MOVPE prepared GaAs surfaces show a striking resemblance with those prepared under MBE conditions [25]. The significance of the RHEED-RAS correlation derived from the MBE experiments becomes apparent. The various reconstructions of GaAs(001) prepared under UHV conditions can unambiguously be identified by RHEED, which probes the long range order. The corresponding RA spectra, being sensitive to the local electronic structure, are very distinct, as well. Therefore, the reconstructions can also be recognised by their RA spectrum. Kamiya *et al* [25] have shown that virtually identical RA spectra can be obtained under MOVPE growth conditions. The local electronic structure which is induced by the formation of surface Ga or As dimers must be the same. This first observation of dimer formation in a non-UHV environment has great importance for the interpretation of MOVPE growth.

Considerable experimental progress has been made in the application of RAS to more realistic MOVPE growth situations by the group of Richter *et al* [11]. They have studied MOVPE growth processes at higher temperatures and with both group III and group V precursors present in the gas phase at H_2 partial pressures around 100 mbar (104 Pa). Figure 4.12 displays the changes observed in the RA spectra when heating a GaAs(001) substrate in a MOVPE reactor. Using the RAS-RHEED correlation mentioned above, the reconstructions can clearly be identified, as indicated in Fig. 4.12 by d(4x4), c(4x4), and (2x4). At intermediate stages, the RA spectra can be convoluted by linear combinations of the c(4x4) and (2x4) RA spectra.

The loss of As occurring during the heating process of Fig. 4.12 can be prevented by adding As-containing precursors into the MOVPE gas flow. The effect of additional AsH_3 at a partial pressure of 0.7 mbar (70 Pa) is shown in Fig. 4.13. At a temperature of 825 K, the (2x4)-type RA spectrum was dominating in the case of pure H_2 flow, whereas now the As-rich c(4x4) surface is still present.

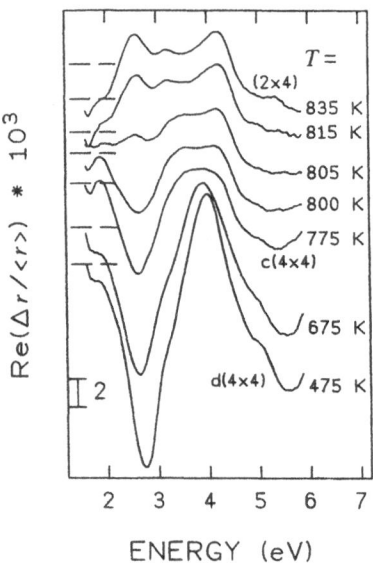

Fig. 4.12. RA spectra of GaAs(001) in a MOVPE reactor at different temperatures and H_2 partial pressure of 100 mbar (10^4 Pa) [11].

Taking a different As precursor, tertiarybutylarsine, at a much lower partial pressure of 0.11 mbar (11 Pa), the loss of As is suppressed more efficiently, as demonstrated in Fig. 4.14. In this case the As-rich c(4x4) RA spectrum is still dominating at 1000 K. The obvious conclusion, which is also corroborated by more indirect growth observations, is that tertiarybutylarsine is a more efficient donor of As than AsH_3. Such results can now be utilised to establish the optimum As stoichiometry at the GaAs(001) surface for growth. From standard

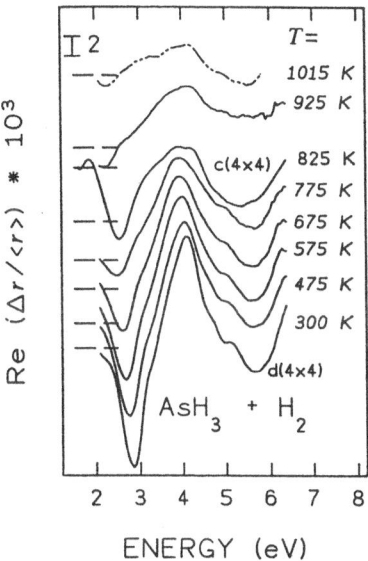

Fig. 4.13. RA spectra of GaAs(001) in a MOVPE reactor at different temperatures and H_2 partial pressure of 100 mbar (10^4 Pa) and AsH_3 partial pressure of 0.7 mbar (70 Pa) [11].

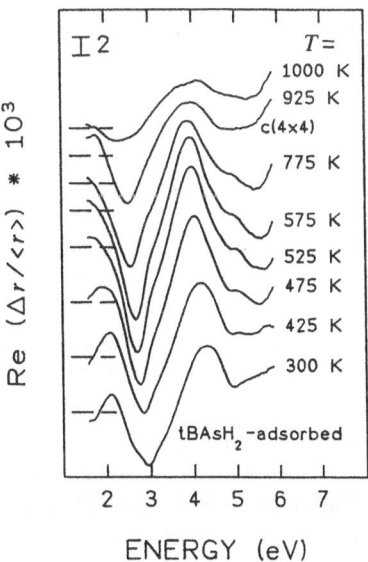

Fig. 4.14. RA spectra of GaAs(001) in a MOVPE reactor at different temperatures and H_2 partial pressure of 100 mbar (10^4 Pa) and tertiarybutylarsine partial pressure of 0.11 mbar (11 Pa) [11].

growth parameters quoted in the literature for MOVPE growth of GaAs, the optimum pre-growth condition is close to the c(4x4) surface.

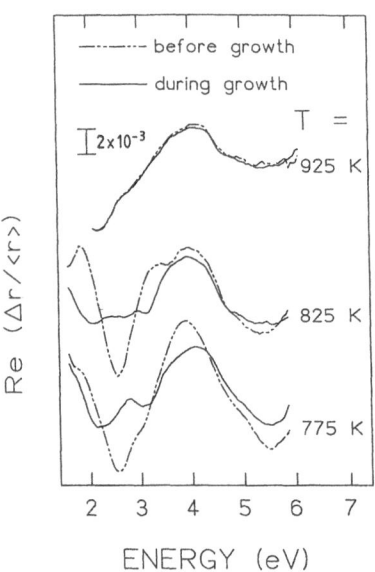

Fig. 4.15. RA spectra of GaAs(001) in a MOVPE reactor at three temperatures and H_2 partial pressure of 100 mbar (10^4 Pa) and AsH_3 partial pressure of 0.7 mbar (70 Pa) without TMG (dashed line), and with a TMG partial pressure of 4.2×10^{-3} mbar (0.42 Pa) (solid line) [11].

Starting growth of GaAs by adding TMG, considerable changes in the RA spectra can occur, compared with the As-stabilised condition. This is illustrated in Fig. 4.15 for three different temperatures. The RA spectra shown, for stationary growth conditions (solid lines), were taken with sampling times for each data point several times longer than the time needed to grow a monolayer of GaAs. The magnitude of change between the spectra before and during growth depends on the degree of As stabilisation of the surface. At higher temperatures, where the surface, before TMG addition, is already close to a (2x4) reconstruction essentially no difference exists between the surface before and after TMG addition. At lower temperatures, where the surface is c(4x4)-like, the changes introduced upon TMG addition are quite strong. They appear especially at those spectral positions where the As dimer transitions are assumed to contribute to the polarisability (2.6 and 4.0 eV). These changes obviously indicate that, as a consequence of additional Ga on the surface, some of the As dimers aligned along the [110] direction are now either destroyed or switch to the [$\bar{1}$10] direction, as would be appropriate for the (2x4) surface.

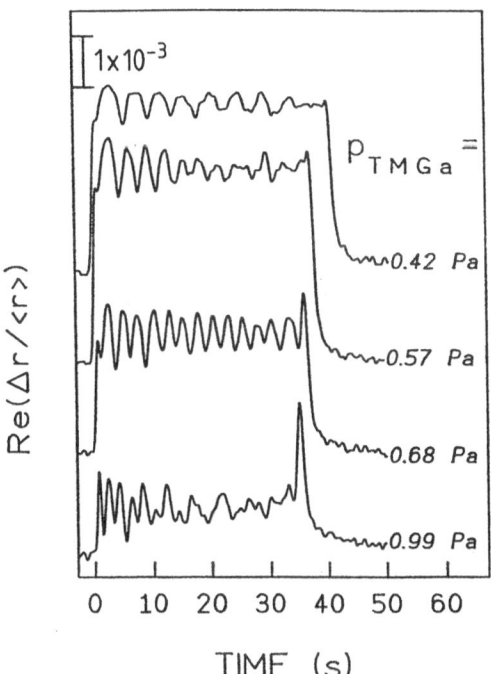

Fig. 4.16. RA signal at a fixed photon energy of 2.6 eV of GaAs(001) and H_2 partial pressure of 100 mbar (10^4 Pa) and AsH_3 partial pressure of 0.7 mbar (70 Pa) when starting growth by adding TMG with different partial pressures between t = 0 s and t = 35 to 40 s [11].

The difference in the RA signals between the pre-growth and stationary growth situations suggests that time resolved studies are feasible for observing growth oscillations similar to those described above for MBE growth. It is also evident from Fig. 4.15 that the time variation (increase or decrease) of the signal will strongly depend on the choice of photon energy. The time dependence monitored at a photon energy of 2.6 eV for GaAs growth at 775 K and a number of different TMG partial pressures is displayed in Fig. 4.16. At t = 0 s, the TMG flux is switched on and coincides with the arrival of TMG at the surface. The rise time, as well as the return time, of the signal after shutting off the TMG flux (at t = 35 to 40 s) are experimentally limited by the time constant of the lock-in amplifier. The magnitude of the rise at t = 0 s corresponds to the difference between the two stationary RA spectra at 775 K in Fig. 4.15, but recorded now for different partial pressures of TMG. The most prominent features occurring after the first rise in Fig. 4.15 are the oscillations. The frequency of the oscillations increases with increasing TMG partial pressure. The period of oscillation corresponds to the growth of one ML of GaAs under these conditions, where growth is transport limited and depends linearly on the TMG partial pressure. Indeed, the growth rates derived from the oscillation periods increase linearly with TMG partial pressure and excellent agreement is found with *ex situ* growth rate evaluation. The RA growth oscillations observed in MOVPE are of better quality than those detected in MBE and clearly underline the strength of this method as an on-line growth monitor.

4.4 Surface Photoabsorption

A technique which is experimentally similar to RAS was developed by Kobayashi and Horikoshi [26] and named surface photoabsorption (SPA). In this case, however, the EM radiation does not hit the sample surface at normal incidence, but at a very shallow angle of typically 70° with respect to the surface normal. This is close to the Brewster angle at which the bulk contribution to the reflected intensity is minimal for *p*-polarised EM radiation (Sect. 1.3). The dependence of the reflectance of *s*- and *p*-polarised EM radiation on the angle of incidence is sketched in Fig. 4.17 for a GaAs substrate and a wavelength of 400 nm.

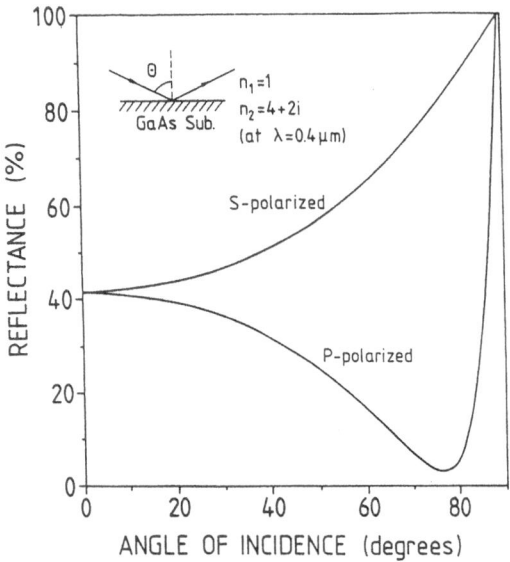

Fig. 4.17. Calculated variation of reflectance upon angle of incidence for *p*- and *s*-polarised light at $\lambda = 400$ nm, using $n = 4 + 2i$ (after [29]).

It can be seen that the reflectance of *p*-polarised light drops to a few percent near the Brewster angle of 76°. One may therefore expect that any modification of the surface which alters the surface reflectivity will reveal itself most markedly in a change of reflectance for *p*-polarised light.

The experimental configuration consists of a light source which can be a laser providing polarised light, or a Xe lamp with monochromator and polariser. The intensity of the reflected light is detected by a Si p-i-n photodiode using a lock-in amplifier. The SPA technique was applied to ALE growth methods which alternately supply the group III and V elements in quantities which allow one atomic layer to be formed upon deposition. Such an approach is feasible in MBE, for example by supplying elemental Ga and As$_4$ molecules alternately, as well as in MOVPE where the constituents are supplied in their gaseous form as, for example, TMG and AsH$_3$. The former method is also called migration-enhanced epitaxy (MEE), while the latter is also termed flow-rate modulation epitaxy by Kobayashi and Horikoshi. According to this growth procedure, the surface is alternately terminated by either Ga or As radicals of the metal-organic compound and AsH$_3$. Consequently, the Authors define their SPA signal as:

$$\frac{\Delta R}{R}\bigg|_{SPA} = \frac{R_{Ga} - R_{As}}{N} \tag{4.16}$$

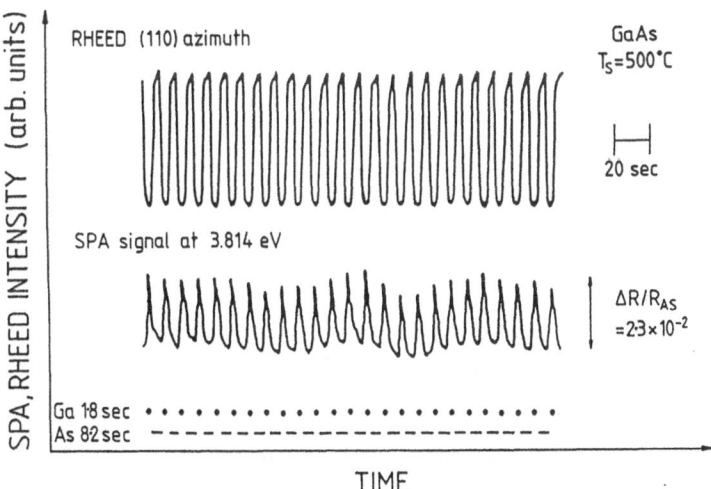

Fig. 4.18. RHEED and SPA intensities for MEE-growth of GaAs. Ga and As MBE cell shutters are opened alternately for 1.8 s and 8.2 s, respectively, where 1.8 s corresponds to a number of Ga atoms equal to the number of surface sites. A He-Cd laser was used ($\hbar\omega = 3.814\,eV$). The azimuth of incidence is [110] for both techniques (after [26]).

Here R_{Ga} and R_{As} are the absolute reflectance intensities of the Ga- and As-terminated surfaces, respectively, and N is a normalisation quantity which varies in the different publications of the group as R_{As} [26-28], R (the average intensity of R_{As} and R_{Ga}) [29], or I, the intensity of the incident light [30].

Considering now MBE-ALE, it is again possible to compare the SPA data with simultaneously recorded RHEED intensities: Fig. 4.18 shows the result [26]. It is apparent that persistent oscillations not only occur in the RHEED intensity, but also in the SPA signal, with an excellent S/N ratio. The difference between the peaks and valleys corresponding to the Ga- and As-terminated surfaces, respectively, is more than 2% of the reflected intensity.

Another As desorption experiment demonstrates that the SPA intensity evolution exactly mirrors the RHEED intensity. Measurements of the spectral dependence of $\Delta R/R_{As}$ (Fig. 4.19) reveal a resonance enhancement near 3.8 eV. Figure 4.19 also shows that such modulation only occurs for p-polarised light. Furthermore, it was also shown that the SPA signal is strongly dependent on the composition of the material, and provides chemical information as well [29].

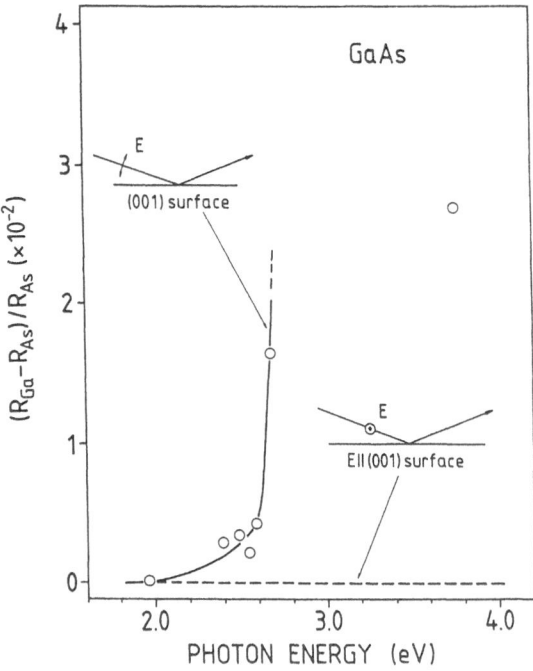

Fig. 4.19. Spectral dependence of $\Delta R/R_{As}$ during MBE-ALE of GaAs using linearly polarised laser light from HeNe (1.959 eV), Ar (2.409 to 2.707 eV), and He-Cd (3.814 eV) lasers (after [26]).

In discussing the origin of the SPA signal, Kobayashi and Horikoshi argue that SPA measures the component of the dielectric tensor parallel to the surface normal since the electric field vector of the p-polarised incident light is almost perpendicular to the surface. However, Hingerl *et al* have clearly pointed out that the contribution to the SPA signal from the dielectric component normal to the surface is negligible and that the major contribution arises from the projection of the dielectric tensor on the line formed by the intersection of the surface and the plane of incidence [31]. Furthermore, Hingerl *et al* also revealed the correlation between SPA and RAS, and demonstrated that their RAS data base can used to calculate SPA spectra (see Sect. 1.3). One advantage of the off-normal SPA measurements is an improved S/N ratio [31]. SPA, RAS and SE are complementary techniques and a simultaneous application can contribute to an improved understanding of surface optics and, thereby, surface chemistry.

References

1. McIntyre, J.D.E., Aspnes, D.E.: Surface Science *24*, 417 (1971)
2. Selci, S., Ciccacci, F., Chiarotti, G., Chiaradia, P., Cricenti, A.: Journal of Vacuum Science and Technology *A 5*, 327 (1987)
3. Aspnes, D.E., Studna, A.A.: Physical Review *B 27*, 985 (1983)
4. Chadi, D.J.: Physical Review *B18*, 1800 (1978)
5. Chiaradia, P., Cricenti, A., Selci, S., Chiarotti, G.: Physical Review Letters *52*, 1145 (1984)
6. Pandey, K.C.: Physical Review Letters *47*, 1913 (1981). *Ibid 49*, 223 (1982)
7. Selci, S., Ciccacci, F., Cricenti, A., Felici, A.C., Goletti, C., Chiaradia,P.: Solid State Communication.*62*, 833 (1987)
8. Cricenti, A., Selci, S., Felici, A.C., Ferrari, L., Gavrilovich, A., Goletti, C., Chiarotti, G.: Journal of Vacuum Science and Technology *A 9*, 1026 (1991)
9. Aspnes, D.E.: Journal of Vacuum Science and Technology *B 3*, 1498 (1985)
10. Acher, O., Koch, S.M., Omnes, F., Defour, M., Razeghi, M., Drevillon, B.: Journal of Applied Physics *68*, 3564 (1990)
11. Richter, W.: Philosophical Transactions of the Royal Society of London A 344, 453 (1993)
12. Paulsson, G., Deppert, K., Jeppesen, S., Jönsson, J., Samuelson, L., Schmidt, P.: Journal of Crystal Growth *111*, 115 (1991)
13. Aspnes, D.E., Harbison, J.P., Studna, A.A, Florez, L.T.: Journal of Vacuum Science and Technology *A 6*, 1327 (1988)
14. Wassermeier, M., Kamiya, I., Aspnes, D.E., Florez, L.T., Harbison, J.P., Petroff, P.M.: Journal of Vacuum Science and Technology *B 9*, 2263 (1991)
15. Studna, A.A., Aspnes, D.E., Florez, L.T., Harbison, J.P., Ryan, R.: Journal of Vacuum Science and Technology *A 7*, 3291 (1989)
16. Aspnes, D.E., Chang, Y.C., Studna, A.A., Florez, L.T., Farrell, H.H., Harbison, J.P.: Physical Review Letters *64*, 192 (1990)
17. Däweritz, L., Hey, R.: Surface Science *236*, 15 (1990)
18. Aspnes, D.E., Harbison, J.P., Studna, A.A., Florez, L.T.: Physical Review *B 59*, 1687 (1987)

102   D. R. T. Zahn

:

19. Harbison, J.J., Aspnes, D.E., Studna, A.A., Florez, L.T., Kelly, M.K.: Applied Physics Letters 52, 2046 (1988)
20. Briones, F., Golmayo, D., Gonzalez, L., deMiguel, J.L.: Japanese Journal of Applied Physics 24, L478 (1985)
21. Scholz, S.M., Müller, A.B., Richter, W., Zahn, D.R.T., Westwood, D.I., Woolf, D.A., Williams, R.H.: Journal of Vacuum Science and Technology B 10, 1710 (1992)
22. Aspnes, D.E.: Thin Solid Films 89, 249 (1982)
23. Kittel, C.: Introduction to Solid State Physics, 5th edition. Wiley, New York 1975, p. 405.
24. Colas, E., Aspnes, D.E., Bhat, R., Studna, A.A., Harbison, J.P., Florez, L.T., Koza, M.A., Keramidas, V.G.: Journal of Crystal Growth 107, 47, 1991
25. Kamiya, I., Aspnes, D.E., Tanaka, H., Florez, L.T., Harbison, J.P., Bhat, R.: Physical Review Letters 68, 627 (1992)
26. Kobayashi, N., Horikoshi, Y.: Japanese Journal of Applied Physics 28, L1880 (1989)
27. Makimoto, T., Yamauchi, Y., Kobayashi, N., Horikoshi,Y.: Japanese Journal of Applied Physics 29, L645 (1990)
28. Kobayashi, N., Horikoshi, Y.: Japanese Journal of Applied Physics 30, L319 (1991)
29. Kobayashi, N., Horikoshi, Y.: Japanese Journal of Applied Physics 29, L702 (1990)
30. Kobayashi, N., Horikoshi, Y.: Japanese Journal of Applied Physics 30, L1443 (1991)
31. Hingerl, K., Aspnes, D.E., Kamiya, I., Florez, L.T.: Applied Physics Letters 63, 885 (1993)

Chapter 5. Raman Spectroscopy

Wolfgang Richter

Institut für Festkörperphysik, Technische Universität Berlin, D-10623 Berlin

5.1 Introduction

Raman spectroscopy (RS) is one of the linear optical techniques widely applied in the analysis of solids, as well as for liquids and gases. It is one of the experimental techniques based on inelastic scattering of light. If the frequency difference between incident and scattered light is in the range below $\hbar\omega = 10^{-7}$ eV, light scattering is termed Rayleigh scattering and Fourier analysis of the detector current yields the change of frequency caused by the inelastic scattering. For larger frequency differences ($\hbar\omega = 10^{-7}$ to 10^{-4} eV), interferometric devices, usually Fabry-Perot interferometers, are used to determine the frequency change. The inelastic light scattering is then termed Brillouin scattering. Finally, Raman scattering describes larger frequency differences (above $\hbar\omega = 10^{-4}$ eV) where grating monochromators can be utilised for frequency analysis of the scattered light.

Since its theoretical prediction in 1923 [1] and its experimental observation in 1928 [2], Raman scattering has been dealing essentially with vibrations as elementary excitations responsible for the light scattering process. The reason for this lies in the fact that the vibrational interaction matrix elements and, consequently, scattering intensities are, in general, much larger than those for other excitations. This nearly exclusive focusing on vibrational properties has relaxed somewhat since the advent of lasers in the seventies as ideal sources for light scattering studies (well defined frequency and wavevector). Other excitations in solids like plasmons, magnons or single electronic transitions then also became experimentally accessible. However, because of their larger scattering signal, vibrational excitations are still the excitations studied in at least 90% of all studies in Raman scattering.

Experimentally, because of the best available light sources (lasers) and light detectors (photomultipliers, silicon based multichannel detectors) light scattering, and Raman scattering specifically, is performed in, or close to, the visible spectral region. Light penetration depths for metal and typical semiconductors in this range are at least 5 to 10 nm, but usually are much larger. For this reason, RS has never really been considered as a surface sensitive technique. However, the penetration depth is not the only way to achieve surface sensitivity. The surface or interface may possess different physical properties, like specific vibrational frequencies or electronic states, as well as macroscopic electric surface fields or strains. All these properties may then serve to differentiate surface properties from bulk properties. Nevertheless, in all such cases a small light penetration depth is useful, since it helps to reduce the bulk signal and will make the surface contribution more easily detectable. On the other hand, a larger penetration depth turns out to be useful in probing buried interfaces and constitutes one advantage of optical techniques, in general, over electron spectroscopies.

A general problem in studying matter with EM radiation is, of course, that the interaction is weak compared to the interaction of matter with electrons. Optical signals generated by a few atomic layers (10^{14} atoms cm^{-2}) in general will be small. This problem can be overcome, however, by utilising resonances of the optical radiation with electronic states, which may increase the sensitivity of the experiment considerably. In Raman scattering, this technique is termed Resonance Raman Scattering (RRS). It will be discussed in detail in this chapter and is probably the main ingredient for the first successful applications of Raman scattering in the investigation of surfaces and interfaces.

5.2   Theory of the Raman Scattering Process

Raman scattering is a form of inelastic light scattering, i.e. a scattering process where an incident photon of frequency, ω_i, causes the creation or annihilation of an elementary excitation in the sample. This is accompanied by the emission of a scattered photon, whose frequency ω_s is shifted with respect to the incident photon. The elementary excitations responsible for such an elastic scattering process may be of different nature, as discussed above. Here, we will mainly focus on optical phonons, because of their importance in the field of interface characterisation. A schematic representation of a scattering experiment is shown in Fig. 5.1.

As is usual in scattering phenomena, a differential cross section, $d\sigma/d\Omega$, is defined as a property characteristic for the material containing N scattering centres:

$$\frac{d\sigma}{d\Omega} = \frac{A}{P_i}\frac{dP_s}{d\Omega} \tag{5.1}$$

The scattering cross section is independent of the experimental configuration and can be related to quantities obtained by calculation. If the number of scattering centres (e.g. the number of molecules in a gas) is known, a differential cross section per scattering centre can be also quoted. For solids it is, in general, more convenient to use a differential cross section normalised to the volume, instead of to the number of scattering centres. The quantity

$$S_{\alpha\beta} = \frac{dP_{s\alpha}}{LP_{i\beta}d\Omega} = \frac{1}{V}\frac{d\sigma}{d\Omega} \tag{5.2}$$

is the one which can be experimentally determined. It represents a differential Raman scattering cross section normalised to the scattering volume. We will, for simplicity, refer to it as the Raman cross section.

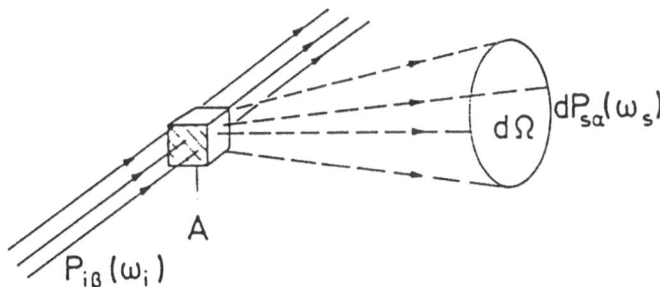

Fig. 5.1. Schematic diagram of a light scattering experiment. $P_{i\beta}(\omega_i)$ is the incident power and $dP_{s\alpha}(\omega_s)$ the scattered power, normalised to the solid angle, $d\Omega$. A denotes the area of the face normal to the incident beam direction.

Many general reviews can be found in the literature which deal with the theoretical aspects of the Raman scattering cross section [3-7]. We will therefore treat in this section only those aspects which are either necessary for a general understanding of the discussion, or those which are specific to interface or surface problems.

The basic aspects of the Raman scattering process, such as energy and wavevector conservation, the general susceptibility approach and the description of light scattering in terms of third order perturbation theory, will be treated first.

The photon-phonon interaction process is mediated by electronic transitions which are then modified by the electron-phonon interaction, since the direct interaction between visible photons and phonons is essentially zero. Finally, selection rules important for the experimental design, and symmetry properties, will be discussed.

5.2.1. Principles of the Raman process

Energy conservation in a scattering process involving one phonon reads

$$\hbar\omega_s = \hbar\omega_i \pm \hbar\omega_j \qquad (5.3)$$

where the plus sign stands for the annihilation of a phonon and the minus sign for phonon creation in the scattering process and ω_j is the phonon frequency. The phonon annihilation process, where the frequency of the scattered light is increased with respect to the incident laser frequency, is called an Anti-Stokes process. In the great majority of experimental investigations, however, the Stokes process, where the frequency of the scattered light is decreased, is utilised. Throughout this chapter only Stokes scattering will be considered.

Similarly, momentum conservation correlates the wavevectors, $\mathbf{k}_i$, of the incident light, and $\mathbf{k}_s$ of the scattered light inside the solid, with the phonon wavevector, $\mathbf{q}_j$:

$$\mathbf{k}_s - \mathbf{k}_i = \mathbf{k} = \pm \mathbf{q}_j \qquad (5.4)$$

From equations 5.3 and 5.4, the scattering wavevector can be related to the possible frequency change in the scattering experiment:

$$|k| = (n_i^2 \omega_i^2 + n_s^2 \omega_s^2 - 2n_i n_s \omega_i \omega_s \cdot \cos\gamma)^{1/2}/c \qquad (5.5)$$

For typical phonon frequencies (100 to 1000 cm^{-1}) and refractive indices (2 to 4), as well as a maximum scattering angle, $\Phi = 180°$, the scattering wavevector is ~10^6 cm^{-1}, i.e. approximately 1% of the BZ. In contrast to x-ray, neutron or electron scattering, therefore, no reciprocal lattice vector can be added in equation 5.4, because the wavevector of visible light is so small. The whole scattering process takes place near the centre of the first BZ. Varying the frequency and scattering angle changes $\mathbf{q}_j$ in the range 10^5 to 10^6 cm^{-1}, typically. Phonons, in general, have only negligible dispersion in this range and thus frequencies measured correspond essentially to $\mathbf{k} = 0$ (i.e. BZ centre). This is not the case if plasmons or coupled phonon-plasma modes are the elementary excitations. As a consequence of diffusion terms in their equation of motion, they show dispersion in this wavevector range and the wave vector is important for the eigenfrequencies quoted [8].

The scattering of light can be described phenomenologically by a generalised susceptibility which is called the transition susceptibility [9]. It connects the incident EM field, $E_\beta(\omega_i)$, with a polarisation, $P_\alpha(\omega_s)$:

$$P_\alpha(\omega_s) = \varepsilon_0 \left\{ \chi_{\alpha\beta}(\omega_i, \omega_s) E_\beta(\omega_i) \right\} \tag{5.6}$$

where α and β denote the polarisation directions of the fields. This general susceptibility can be expanded in a Taylor series as a function of phonon normal coordinates, Q_j, phonon wavevectors, q_j, and additional external perturbations such as electric fields, E_γ, or strains [6]:

$$\chi_{\alpha\beta}(\omega_i, \omega_s) = \chi_{\alpha\beta}(\omega_i, \omega_i)$$

$$+ \sum_j Q_j \cdot \left(\frac{\partial \chi_{\alpha\beta}(\omega_i, \omega_s)}{\partial Q_j} \right)$$

$$+ \sum_j Q_j \cdot q_j \cdot \left(\frac{\partial^2 \chi_{\alpha\beta}(\omega_i, \omega_s)}{\partial Q_j \partial q_j} \right)$$

$$+ \sum_j \sum_\gamma Q_j \cdot E_\gamma \cdot \left(\frac{\partial^2 \chi_{\alpha\beta}(\omega_i, \omega_s)}{\partial Q_j \partial E_\gamma} \right)$$

$$+ \sum_j \sum_\gamma Q_j \cdot q_j \cdot E_\gamma \cdot \left(\frac{\partial^3 \chi_{\alpha\beta}(\omega_i, \omega_s)}{\partial Q_j \partial q_j \partial E_\gamma} \right)$$

$$+ \sum_j \sum_{j'} Q_j \cdot Q_{j'} \cdot \tfrac{1}{2} \left(\frac{\partial^2 \chi_{\alpha\beta}(\omega_i, \omega_s)}{\partial Q_j \partial Q_{j'}} \right)$$

$$+ \ ....$$

$$\tag{5.7}$$

The right hand side of equation 5.7 describes terms of different order in the scattering process. The first term represents the susceptibility of the static crystal and gives elastic scattering (Sect. 1.3, and Ch. 3 and 4), the second term 1-phonon Raman scattering, the terms three to five describe 1-phonon Raman scattering induced by a finite wavevector or electric field and, finally, the last term represents scattering by two phonons, indexed j and j'. This description in terms of macroscopic derivatives does not, of course, give any physical insight into the scattering process, but it is useful for classifying the different possible processes and, moreover, it allows the derivation of selection rules for the transition susceptibility, which are called Raman tensors and are useful in designing the proper experimental configuration.

The Raman cross section can also be expressed directly in terms of the transition susceptibilities. For 1-phonon scattering (terms 2, 3, 4 and 5 in equation 5.7) equation 5.2 becomes [6]

$$S_{\alpha\beta} = \frac{dP_{s\alpha}}{LP_{i\beta}d\Omega} = \frac{1}{V}\frac{d\sigma}{d\Omega} = \frac{\left|\chi_{\alpha\beta}^{j}(\omega_i,\omega_s)\right|^2 V\omega_S^4}{c^4} \qquad (5.8)$$

In the usual Raman experiment the transition susceptibilities connect two electromagnetic fields with photon energies typically in the range from 1 to 4 eV. Electronic transitions dominate the electric susceptibility in this energy range. The derivatives in equation 5.7 are thus derivatives of the electronic susceptibility with respect to lattice deformations described by phonon normal coordinates. The phonons modulate the electric susceptibility by means of the electron-phonon interaction. This fact is more clearly expressed in the microscopic quantum mechanical time-dependent perturbation theory of the Raman scattering process. It yields, for the most resonant term of the process depicted in Fig. 5.2 [11]:

$$\chi_{\alpha\beta}^{j} = \frac{e^2}{m_0^2\omega_s^2 V}\sum_{l,m}\frac{<o|p_\alpha|m><m|H_{EL}|l><l|p_\beta|o>}{(E_m-\hbar\omega_s)(E_l-\hbar\omega_j)} \qquad (5.9)$$

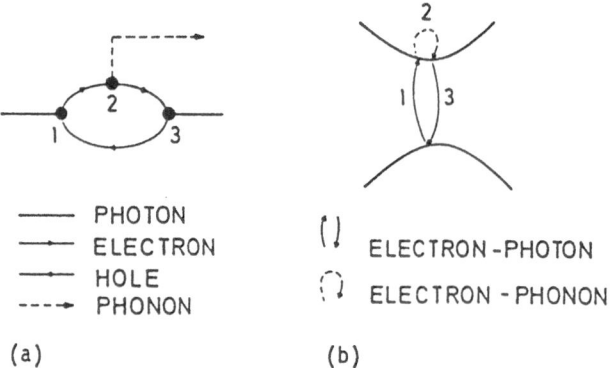

(a)   (b)

PHOTON
ELECTRON
HOLE
PHONON

ELECTRON-PHOTON

ELECTRON - PHONON

Fig. 5.2. Diagram of the Raman scattering process and corresponding transitions in terms of the electronic band structure from a filled valence band to an empty conduction band. Three interactions are displayed: (1) the dipole interaction by which the incident photon creates an electron-hole pair, (2) the electron-phonon interaction which scatters the electron-hole pair from state l to state m and (3) the dipole interaction which, by recombination, leads to the emission of a scattered photon.

In this three-step process it can be seen that, as well as the vector components of the dipole operators, the electron-phonon interaction, H_{EL}, is involved. This is additional information obtained in Raman scattering experiments, as compared to the susceptibility measurements of Ch. 3 and 4 (ellipsometry, reflectivity). The second important point to note from equation 5.9 is the resonance behaviour expressed by the denominators, when the photon energies are close to electron-hole pair energies. This observation constitutes essentially one of the secrets of sensitive Raman scattering experiments, where the laser frequency should be chosen such that it is in close coincidence with points of high density of states for electronic transition energies. It also demonstrates the connection between the Raman cross section and the normal susceptibility which, in second order perturbation theory, is written as:

$$\chi_{\alpha\beta}(\omega) = \frac{e^2}{m_0^2 \omega^2 V} \sum_l \frac{<o|p_\alpha|l><l|p_\beta|o>}{(E_l - \hbar\omega)} \qquad (5.10)$$

This allows, within certain approximations, the Raman cross section to be expressed as a function of derivatives of the susceptibility with respect to frequency [10]. Such relations are extremely useful, not only estimating resonance enhancement with the help of susceptibility data derived, for example, from ellipsometric measurements, but also for predicting resonances and, as a consequence, enabling experimental parameters, such as the exciting laser frequency, to be optimised.

5.2.2 Scattering Mechanism

Two different properties of the electronic states and the electronic eigenfunctions can be modulated by the phonons, giving rise to the electron-phonon interaction matrix element in equation 5.9. Firstly, the electron energy eigenvalues may be modified and this type of interaction is essentially responsible for the fact that the scattering cross section can be related to frequency derivatives of the susceptibility. Secondly, the phonon may modify the eigenfunctions, which may then be described by an admixture of states of adjacent bands. The latter effect is especially noticeable if two bands (valence or conduction) are close in energy. This is found, for example, in semiconductors of diamond and zincblende structure, where there are spin-orbit-split valence bands.

The phonon property which causes these modifications on the electronic states may be either the lattice deformation created by the presence of the phonon, or a macroscopic electric field connected with the phonon. The former case is termed a deformation potential interaction, and the latter case, a Fröhlich interaction. Deformation potential scattering is the dynamical analogue of the piezomodulation of the electronic band structure which has been extensively discussed by Kane [11]. For a static strain, deformation potentials have been defined according to the symmetry of the electronic bands and the strain symmetry. Within the adiabatic approximation, these deformation potentials can be used equivalently in the description of deformation potential scattering by phonons.

Optical phonons in crystals which do not have a centre of symmetry may have both a non-zero susceptibility derivative (Raman active), and a dynamic electric dipole moment (IR active). This is, for example, the case for optical phonons in the zincblende crystal structure (III-V and II-VI semiconductors). For longitudinal phonons, this electric dipole leads to a macroscopic electric polarisation field, which vanishes for transverse phonons. This macroscopic electric field acts as an additional restoring force and leads to a frequency increase of the longitudinal optical (LO) phonons. Moreover, the corresponding potential can induce electronic transitions, leading to an additional Raman scattering channel. This scattering mechanism is called the Fröhlich interaction, after its first discussion by Fröhlich in connection with superconductivity [12]. For LO phonons of very small wavevector ($q_j \approx 0$), this mechanism is small since the contributions from electrons and holes compensate one another. However, for finite q_j values, this compensation is lifted and huge scattering cross sections may be obtained, especially when close to resonance with electronic states. This scattering process, also called Electric-Field-Induced Raman Scattering (EFIRS), is the basis for determining Schottky barrier heights at interfaces of zincblende-type semiconductors by Raman scattering [13].

5.2.3 Selection Rules

The conditions, arising from crystal symmetry, on the various tensor components of the transition susceptibilities are termed *Raman selection rules*. These conditions, most importantly, predict which tensor components are zero, and thus which incident and scattered light polarisations will give zero scattering intensity by symmetry. They serve, therefore, as important guidelines for designing experiments.

Group theory provides the tools for determining the zero tensor components, because the irreducible representations describing the phonon property (eigenvector or macroscopic field) must transform like those representations describing the properties of second rank tensors [14]. For 1-phonon scattering, these selection rules have been determined for all crystal classes. They are called *Raman tensors* and can be found, for example, in Refs. 15 and 16.

5.3 Experiment

In this section, a few remarks on Raman scattering from surfaces and interfaces will be made. For more general aspects of the RS experiment, the reader is referred to the literature [17]. The equipment available nowadays for RS, i.e. lasers, monochromators and detectors, is both sophisticated and highly developed. It can be used for interface studies without any modification. The main concern should be the selection of a laser with spectral lines capable of establishing resonance conditions, together with the choice of single or multichannel detection.

The largest effort, however, is concerned with adapting this equipment to the UHV chamber. In order to obtain sufficient sensitivity, the aperture of the collecting optics should be as high as possible. Thus, the sample should be close to the collecting lens. This might be a problem in standard UHV chambers, because the manipulator may not be able to move the sample close enough to the viewports. As shown in Fig. 5.3, a possible solution is a re-entrant window, which is located some cm inside the UHV chamber. This allows the collecting lens to be placed quite close to the sample without modifying the working radius of the manipulator.

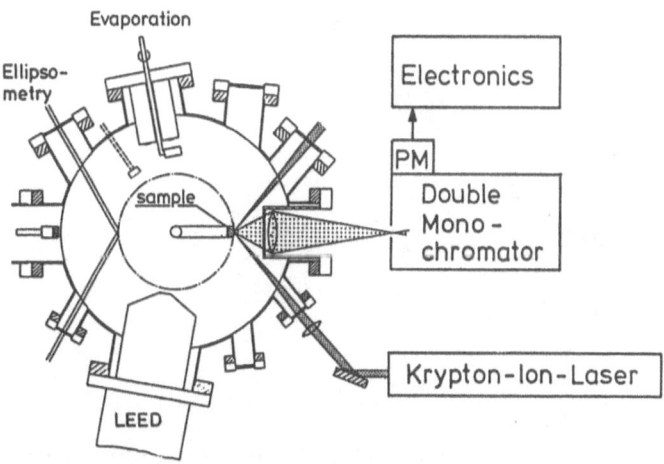

Fig. 5.3. RS configuration for *in situ* investigations under UHV conditions. The special window arrangement allows a sufficient aperture for the scattered light collecting optics.

An alternate design, which requires a specially constructed UHV chamber, allows direct integration of the optics into an MBE set-up which, when combined with a Raman spectrometer with multichannel detection, allows on-line growth monitoring, as shown in Fig. 5.4. Most RS experiments use CW lasers of relatively low power (up to 100 mW). The incident light power is essentially limited by the requirement that the scattering should not influence sample properties, e.g. by generating too many photo carriers, by heating, or by thermal desorption of adsorbed overlayers. Such possible perturbations by the laser source have to be checked in each case, for example by varying the laser power.

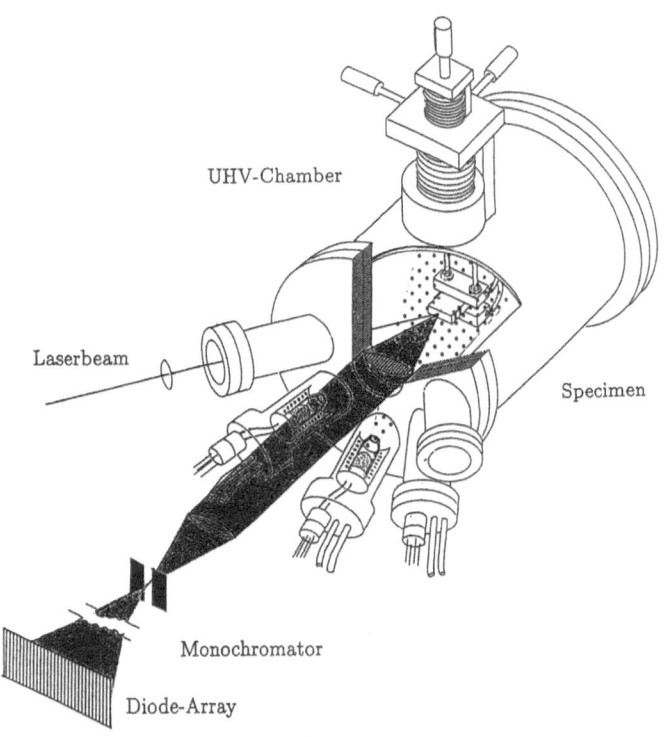

Fig. 5.4. RS configuration for *in situ* on-line MBE monitoring, when a multichannel Raman spectrometer is included, as indicated schematically [18].

Few RS measurements have been performed with pulsed lasers. A number of problems may arise: (i) the high peak powers may significantly modify the sample properties and thus RS may not be a non-perturbative probe any more, (ii) the duty factor of pulsed laser is, in general, small and extensive signal averaging and a low S/N ratio may result, (ii) because of the high peak power excitation, the signal pulses may contain more than one photon and the usual photon counting electronics may have to be modified.

A common choice with respect to lasers for the CW RS experiment are Ar- and Kr-ion lasers, which offer the selection of several lines and, in addition, are very stable and reliable sources. For more accurate resonance tuning, a dye laser pumped by one of the ion lasers is desirable.

The choice of the monochromator system depends essentially on the choice of detector. If this is a photomultiplier, then an additive double monochromator may be applied. Narrow slits provide the resolution and contrast necessary to discriminate against the stray light of the laser. If a multichannel detector is chosen, the monochromator has to transmit a broader spectral range and additional contrast is required. In such a case, a subtractive double monochromator for stray light reduction is usually coupled with a single monochromator generating the dispersion. Both types of monochromator are shown in Fig. 5.5.

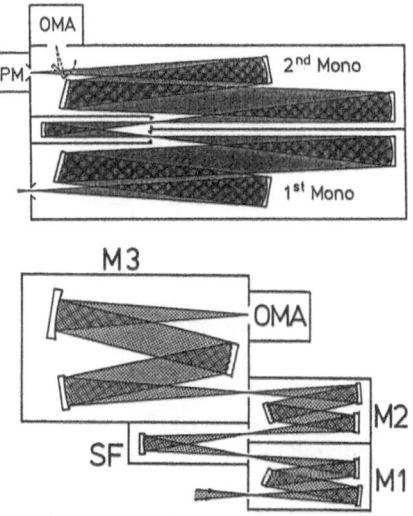

Fig. 5.5. Possible monochromators for the RS experiment: (top) additive double monochromator for use with a photomultiplier, and (bottom) triple monochromator consisting of a subtractive double monochromator and a dispersion-generating third monochromator, for use with a multichannel detector (OMA).

The choice between a photomultiplier- or a multichannel-based system, with respect to performance, will generally be easy, because of the multichannel advantage of the latter. A multichannel-based system may even allow real-time investigations. However, in order to obtain the same detectivity as a photomultiplier system, a careful selection of multichannel detectors has to be made and, in general, the price of such a Raman system will be considerably higher than a photomultiplier-based system.

5.4 Overlayers with Abrupt Interfaces

Abrupt interfaces are the simplest type of a heterostructure interface. Interest in these focuses, firstly, on the overlayer. The structure (which might be different from the bulk structure), the electronic and lattice dynamical properties of the overlayer, and their development with thickness, are the questions to be answered. The influence of the overlayer on the substrate, and the modification of interface states giving rise to band bending is also of interest. The latter question will be discussed in Sect. 5.6.

Raman scattering mainly gives information on the lattice dynamical properties of the overlayer. This will be the main topic of the following sections. However, the phonon properties are essentially determined by the structural properties of the layer, and thus provide an additional test for the structure determination, usually performed by LEED. In addition structural changes manifest themselves quite significantly in the Raman spectra and therefore Raman scattering is well suited for monitoring *in situ* structural phase transitions. The scattering cross section, and especially its variation with excitation frequency, on the other hand, depends strongly on the electronic state properties, as can be seen in equation 5.9. This property might open a new experimental channel for studying the electronic structure of thin overlayers, in addition to the standard methods based on photoemission or STM.

5.4.1 Antimony Overlayers on III-V Semiconductors

Antimony overlayers on III-V semiconductors are amongst the oldest examples of well-defined overlayers with abrupt interfaces. Most investigations have been concerned with GaAs(110) and InP(110) as substrates. A special feature of the group V overlayer systems is that they form a well-ordered first monolayer, whereas the growth mode is different for larger thicknesses. We will therefore discuss Sb monolayer properties first, and then deal with thicker overlayers.

5.4.1.1 Antimony Monolayers

Monolayers have been found to adsorb in an epitaxial manner, as shown in Fig. 5.6. Their electronic structure has been the subject of many investigations by normal [21-24] and inverse photoelectron spectroscopy [25], as well as tunneling microscopy [26, 27]. First investigations of lattice dynamic properties in the acoustic phonon region were performed by He atom scattering [28].

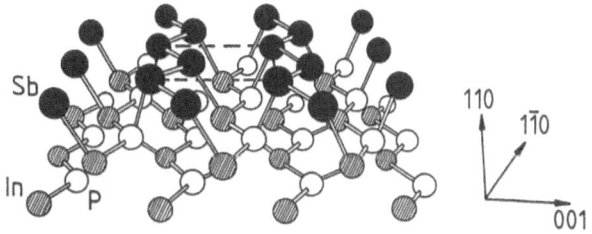

Fig. 5.6. Schematic view of the Epitaxial Continued Layer Structure (ECLS) of 1 ML of Sb on GaAs(110) 1x1 [19, 20]. The dashed line indicates the surface unit cell.

Optical surface phonons were successfully detected for the first time by RS [28], as shown in Fig. 5.7, where the development of Raman scattering peaks of the Sb monolayer on top of an InP(110) substrate is shown.

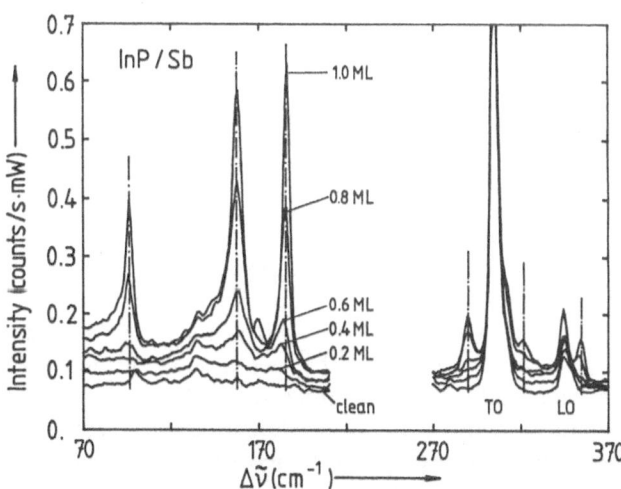

Fig. 5.7. RS peaks of Sb on InP(110) for different coverages between 0 and 1 ML. The epitaxial growth leads to vibrational peaks which differ from those of bulk Sb and the substrate [29].

Proof that these peaks are due to vibrational modes of the Sb monolayer was given by the thickness dependence, since the peak intensities level off at 1 ML coverage (Fig. 5.8). Further evidence has come from theoretical calculations [30, 31].

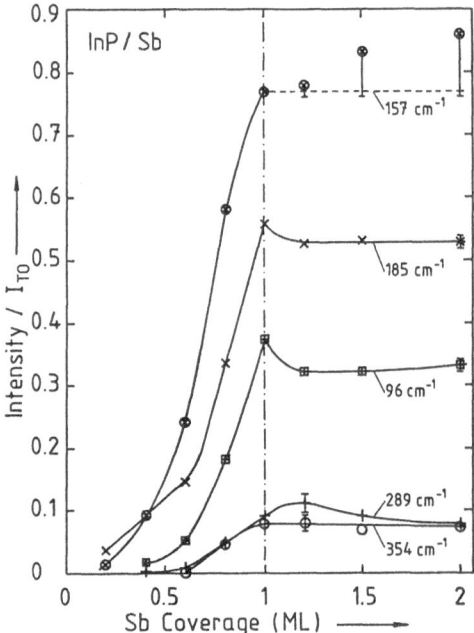

Fig. 5.8. RS intensities of optical surface phonons of Sb on InP(110), as function of coverage [29].

Altogether six peaks are observable on first inspection. However, polarisation analysis of the scattered light in different configuration (Fig. 5.9) shows that there are altogether 8 different scattering frequencies. The assignment of these scattering peaks to certain surface phonons starts with a normal coordinate analysis of the possible vibrations. The fact that, altogether, 8 phonons are observed is not compatible with assuming just 2 Sb atoms per unit cell. This would give 4 optical modes plus 2 acoustic ones in 2 dimensions. In order to obtain a larger number of modes, the topmost substrate layer has also to be involved. With 4 atoms per unit cell (2 Sb, 1 In, 1 P) in 3 dimensions, 12 modes are obtained altogether, 9 of which are optical and 3 acoustic. The 9 optical ones are sufficient to describe the experimental scattering data with 8 eigenfrequencies. Assignment of normal coordinates to particular frequencies is difficult, however. A' and A" modes may be distinguished on the basis of Raman tensor selection rules for the low symmetry system (C_s). This allows the assignment of the peaks at 96, 157, 185, 289, 321 and 354 cm^{-1} to A' symmetry, and the peaks at 161 and 290 cm^{-1} to A" symmetry. In addition, one group of peaks appears in the range of the substrate TO and LO phonons, while the other group shows up in the frequency range of bulk Sb lattice frequencies. It can be concluded that the former mainly involve bonds from the substrate atoms (InP), while the latter are dominated by Sb-Sb bonds. Lattice dynamical calculations performed recently for Sb on GaAs(110) [30, 31] provide further evidence for the tentative assignment given in Ref. 32.

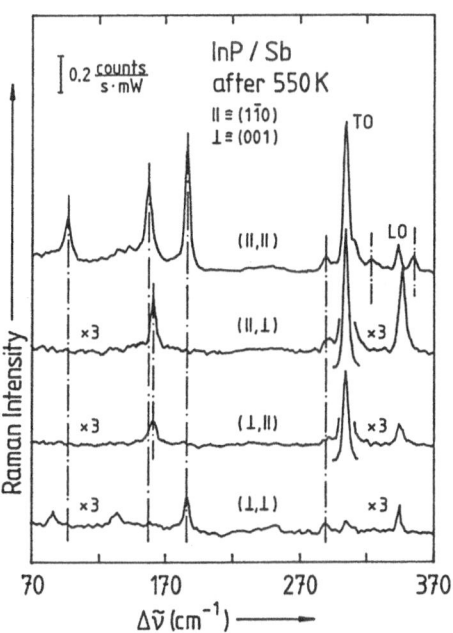

Fig. 5.9. Raman spectra of 1 ML Sb on InP(110), measured in different polarisation configurations [29].

Resonance measurement, i.e. the measurement of the scattering cross section as a function of incident laser frequency, have been also performed for these vibrations (Figs 5.10, 5.11) [32]. Resonance enhancement is observed for energies which differ from those of bulk Sb and the TO resonance in GaAs and InP. This indicates quite strongly that the resonance involves electronic states in the 2-dimensional bandstructure of the Sb overlayer. Indeed, the resonance energies observed correlate quite well with those derived from bandstructure calculations [19, 20, 33-35], and from STM measurements [26, 27].

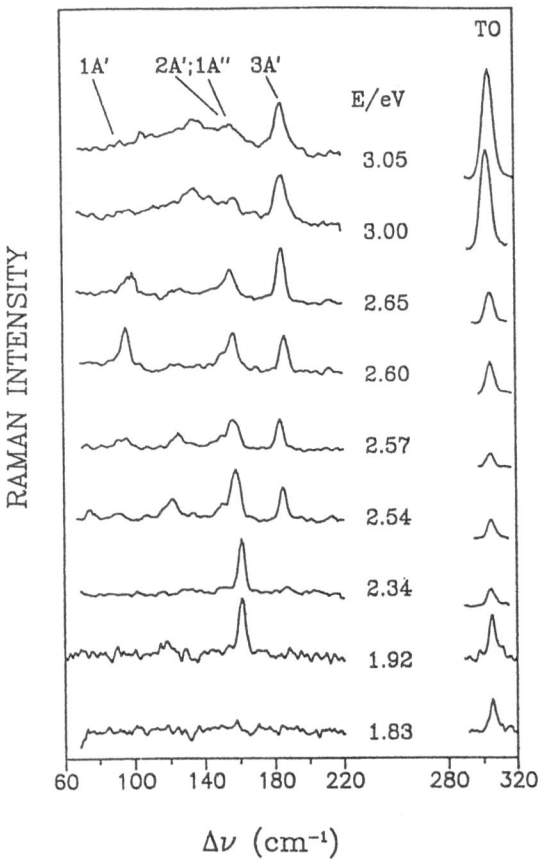

Fig. 5.10. Raman spectra of an Sb monolayer on InP(110) taken with different laser lines of a Kr ion laser. As well as the InP substrate TO phonon, three monolayer modes of symmetry A' and one mode of symmetry A" are shown [32].

5.4.1.2 Thicker Sb Overlayers

Depositing more than 1 ML at RT on (110) surfaces of different III-V semiconductors leads to similar pictures in all cases studied so far. Fig. 5.12 gives an example for Sb on GaAs(110). A broad feature appears for coverages below 15 ML which is typical of amorphous Sb [38]. The scattering spectra of amorphous materials essentially reflect the phonon density of states throughout the BZ. Accordingly, a broad band is observed which in the case of Sb is centred around 135 cm-1 with a half width of 30 to 40 cm-1.

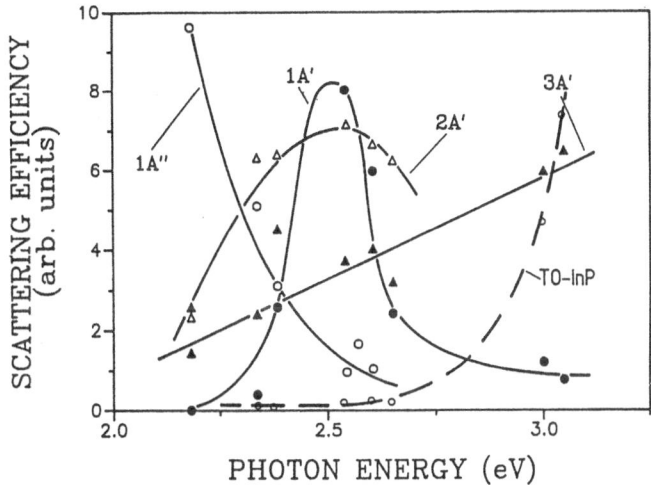

Fig. 5.11. Normalised RS intensities of the 1A' mode (96 cm^{-1}), 2A' (157 cm^{-1}), 3A' (185 cm^{-1}) and the 1A" mode (161 cm^{-1}) for the InP(110) 1x1-Sb overlayer system (solid lines). The resonance curve of the InP bulk TO phonon [36] (dashed line) differs from that of the monolayer scattering peaks [37].

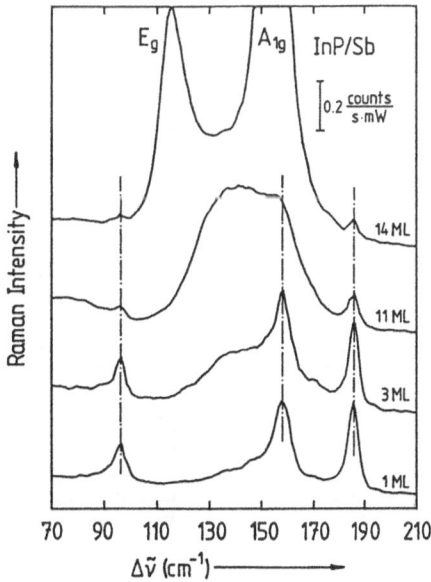

Fig. 5.12. RS spectra of Sb on InP(110) for coverages above 1 ML. The broad structure indicates amorphous Sb while the two strong peaks labelled E$_g$ and A$_{1g}$ at larger coverages (14 ML) correspond to the rhombohedral structure of bulk Sb [29].

Around 15 ML, the broad structure changes rather abruptly into two sharp peaks. Their frequencies are already close to the frequencies of the E_g and A_{1g} modes of bulk Sb. This is evidence for a structural phase transition from amorphous to rhombohedral Sb. It should be noted that this phase transition was not observed with LEED, because the crystallite size is smaller than the necessary coherence area required. With increasing layer thickness, the frequencies approach the exact bulk values [39]. The somewhat different frequencies at lower coverages indicate that the layers are strained and relax with larger layer thickness. Stress values estimated from the compliances of bulk Sb and the known Grüneisen parameter for the phonons [40] amount to a few kbar. The observation of surface phonon peaks up to the highest coverages, in Fig. 5.12, indicates also that the epitaxial structure of the first monolayer remains essentially intact, even after crystallisation of the overlayer.

The growth mode just described is only typical for RT deposition. At low temperatures (80K) the phase transition is suppressed and the structure remains amorphous [42]. On the other hand, if thin amorphous structures with thickness between 2 and 10 ML (deposited at RT) are annealed at temperatures between 350 and 500K, another metastable crystalline modification with biaxial crystal symmetry is formed [41]. As a consequence, three modes different in frequency are observed (Fig. 5.13).

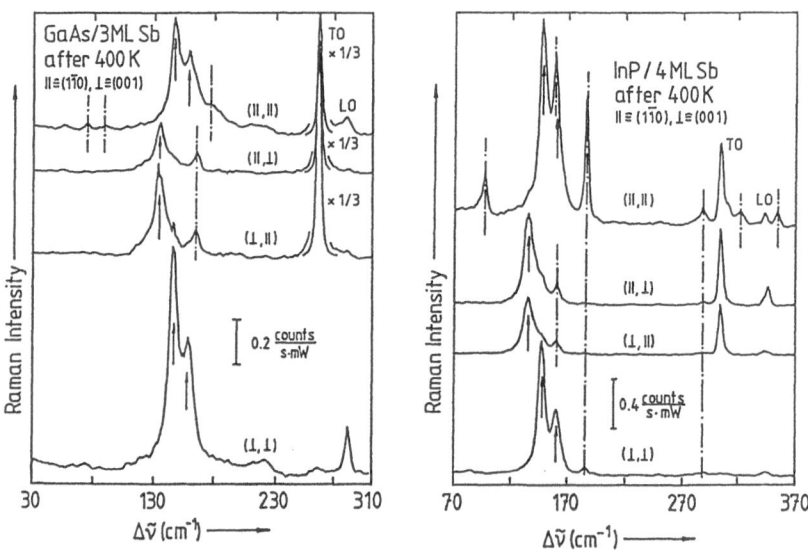

Fig. 5.13. Raman spectra of Sb on GaAs(110) (left) and InP(110) (right) after annealing to 400K, taken for different polarisation configurations. The arrows mark the peaks due to lattice vibrational modes of the orthorhombic Sb modification [41].

5.4.2 Bismuth Overlayers on III-V Semiconductors

The interface of Bi on III-V semiconductors is another example of an ideally abrupt interface. However, the Bi monolayers are not perfectly ordered [43], and this might be one reason why, up to now, no clear indication of monolayer modes has been observed (see Fig. 5.14, bottom). Another reason might be, of course, that experiments have not been performed at resonance.

For thicker coverages of a few monolayer, crystalline phases have been found with substrate-induced structure [41]. The structure is revealed as orthorhombic by LEED, and in RS studies two peaks at 74 and 90 cm^{-1} are observable from 3 ML on (Fig. 5.14, top). This phase corresponds to the orthorhombic Bi phase obtained by annealing of a few monolayers at higher temperatures.

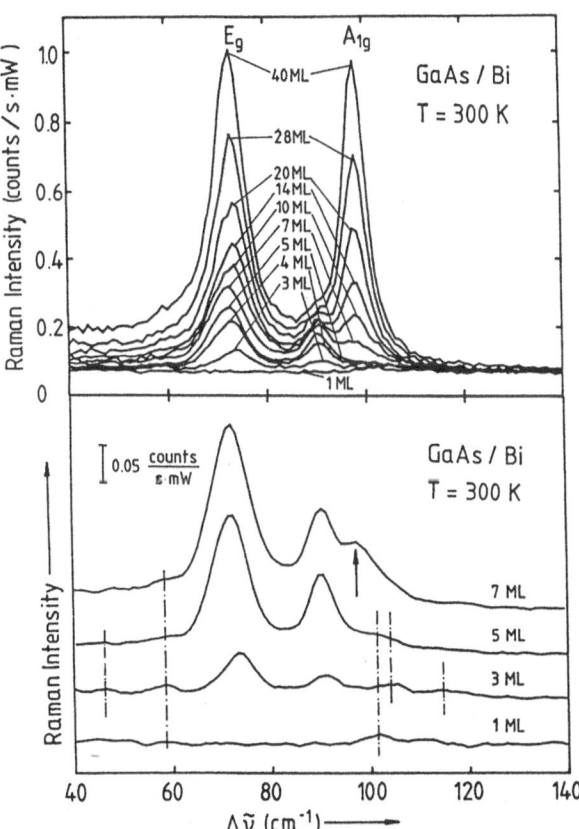

Fig. 5.14. RS of Bi overlayers up to 7 ML (bottom), and up to 40 ML (top), on GaAs(110). The peak shifts between 7 and 20 ML deposition mirror the transition from a orthorhombic surface stabilised structure to the bulk rhombohedral structure [43].

Around 8 ML, a second set of peaks at 72 and 98 cm^{-1} nearly coincident with those of bulk Bi evolves, and a trigonal LEED pattern indicates that the rhombohedral bulk structure has now been formed. The evolution of the Bi Raman spectra with increasing thickness shows that, as in the case of Sb, two structural phases are formed, namely orthorhombic and rhombohedral. For Bi, however, both phases are obtained for RT deposition, while Sb requires higher temperatures. RT deposition of Sb produces an amorphous phase, instead of the orthorhombic phase obtained with Bi.

5.4.3 Germanium on GaAs

Both Ge and GaAs crystallise in the diamond/zincblende structure. Their lattice constants are quite similar (Ge: 0.5658 nm, GaAs: 0.5653 nm). Therefore, ideal epitaxial growth of these two pseudomorphic structures should be possible. Indeed *in situ* RS measurements show the TO phonon of crystalline Ge [44] quite clearly (Fig. 5.15), if deposition takes places at 675K. At low coverages, the TO line

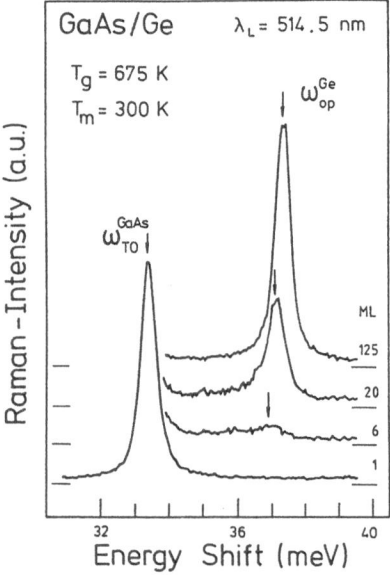

Fig. 5.15. RS of Ge on GaAs(110) heterostructures, for different coverages of Ge. At the substrate temperature of 675K, the Ge overlayer grows epitaxially [44].

is broadened due to relaxed momentum conservation in the thin layer. In addition, as was discussed already for Sb overlayers, the frequency is shifted in comparison to the bulk Ge TO frequency, due to strain in the thin layer. Above 25 ML, however, the linewidth, as well as the frequency, approaches the bulk Ge values. For lower temperatures, polycrystalline or even amorphous growth of Ge was observed. This was clearly indicated by a broad feature in the Raman spectra representing essentially the phonon density of states due to wavevector non-conservation [44].

5.4.4 Arsenic on Si(111)

Arsenic layers on Si have received considerable attention, as they provide the possibility of protecting UHV-prepared Si surfaces against exposure to air, for example, for transportation purposes. This seal, usually called a cap, can then be removed at moderate temperatures (550°C) in a different UHV chamber, thereby recovering the original Si surface for further study.

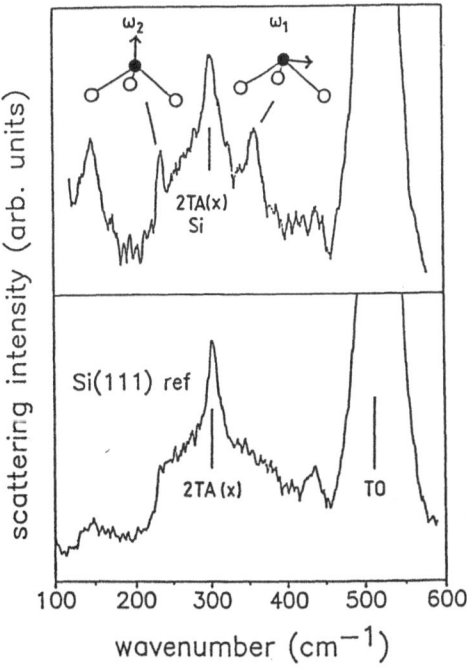

Fig. 5.16. Raman spectrum of 1 ML As on Si(111) (top) compared to that of Si(111). For the two As-induced features, the ω_1-vibration is assigned to the doubly-degenerate E-mode, and the ω_2-vibration to the A-mode [46].

The structure of the thick as-grown cap consist mainly of amorphous As. This amorphous layer is removed at temperatures around 380°C. An ordered monolayer of As remains on the surface. On Si(111), the As atoms take Si atomic positions

[45], and the structure belongs to the so-called Epitaxial Continued Layer Structures (ECLS). The surface unit cell contains one As atom. The geometry allows three eigenmodes, two of which are doubly degenerated within the xy-plane (ω_1) and one single mode with displacements along z (ω_2). Two peaks with different frequencies are indeed observed, as can be seen in Fig. 5.16. There, the RS spectrum of Si(111) covered with a monolayer is compared to the spectrum of an uncovered Si(111) surface. As well as the two main Si structures, the TO phonon at 520 cm^{-1} and the 2TA(X) 2-phonon peak around 300 cm^{-1}, two new peaks can clearly be observed on the As-covered surface. The assignment of the eigenmodes (shown in the figure) was made using polarisation selection rules [46]. Evidence that these two peaks are due to the As monolayer can be also obtained by removing the monolayer at higher temperatures. This is shown in Fig. 5.17, where the two peaks disappear at higher temperatures, to leave the spectrum of clean Si(111).

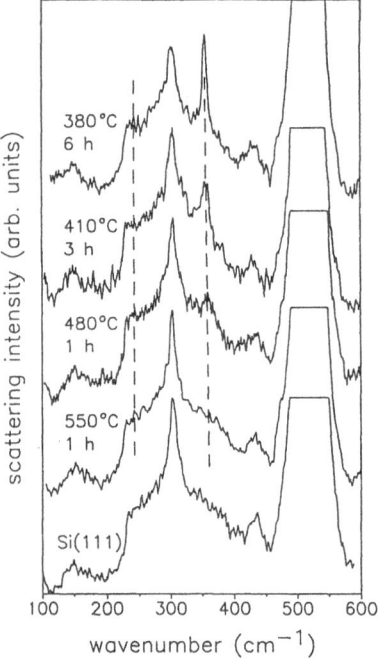

Fig. 5.17. RS spectra taken after annealing at higher temperatures. The two additional features attributed to the As monolayer disappear at the highest annealing temperature (550°C). The spectrum is then identical to that of Si(111), shown at the bottom for comparison [47].

Finally, Fig. 5.18 shows the importance of resonance behaviour for the detection of such monolayer modes. The E-mode of the As monolayer is clearly enhanced for higher photon excitation energies, especially in comparison to the behaviour of the Si 2TA(X) or the Si TO phonon. This indicates that there exists a gap at these energies in the surface electronic transitions of the As monolayer on Si.

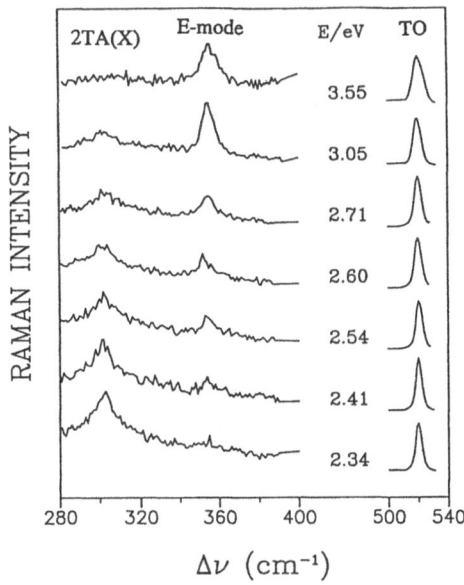

Fig. 5.18. Raman spectra of the E-mode at 365 cm^{-1} of the As monolayer, for different photon energies of the exciting laser [46].

5.5 Reactive Overlayers

The majority of interfaces between different materials are non-abrupt. This occurs either through diffusion processes, quite often enhanced by relatively high growth temperatures, or because overlayer elemental species are highly reactive with the substrate. While diffusion in most cases can be limited to a few monolayers, reactions occur, especially at high temperature where they may modify the interface region considerably. RS can monitor these events because the vibrational excitations will be changed in that region. Identification of new phases, created at the interface, is often possible with the help of calibration spectra. In studying such a system, both the development of layer properties with layer thickness, and also the structure of the interface, are important problems.

5.5.1 Reactive Growth of III-V Semiconductors

Little *in situ* work has been performed on these systems, even though they are of major importance in the area of semiconductor technology and devices. Recently, however, nice examples which show the potential of *in situ* and on-line monitoring by RS have appeared, dealing with the growth of III-V semiconductor layers *via* the reaction of group III elements (In, Al, Ga) on group V substrates (Sb, Bi). One prominent example is the growth of InSb on Sb, as shown in Fig. 5.19. The clean Sb(001) surface (cleavage plane) shows the two bulk Sb phonon peaks at 113 and 150 cm⁻¹.

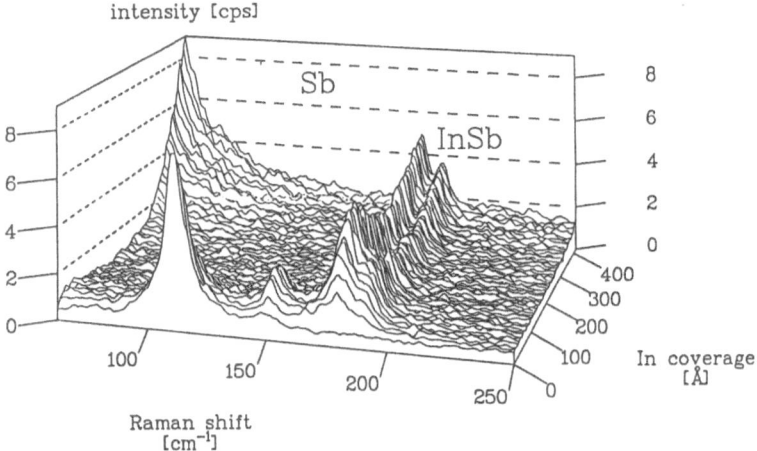

Fig. 5.19. On-line *in situ* RS of InSb grown epitaxially at room temperature by depositing In on Sb(111) [48].

With increasing nominal In coverage, the TO and LO phonon structures of InSb evolve. The line widths and selection rules of both phonons correspond to that of bulk single crystalline InSb with a (111) surface, as expected. It can be also seen in Fig. 5.19 that, simultaneously, the Sb phonon scattering peaks are disappearing, which indicates layer growth. The intensity variation can be simulated with a two layer system, which allows the layer thickness to be determined [48].

5.5.2 Growth of II-VI Semiconductors on III-V Substrates

The growth of II-VI semiconductors on III-V semiconductors constitutes an important technological problem, since these interfaces are highly reacting. Depending on temperature and the elemental flux, different amounts of III-VI compounds are formed, with the group V atoms being liberated in elemental form. One of the earliest example of Raman studies of these systems is that of MBE-grown CdTe on InSb(001) [49]. Fig. 5.20 shows the Raman spectra of two samples which were nominally deposited with 30 nm of CdTe. The use of a single CdTe source leads to a deficiency of Cd resulting in the formation of In_2Te_3 and elemental Sb. This can be overcome using a second source to provide additional Cd, which leads to the formation of CdTe (Fig. 5.20b). Similarly, lowering the temperature reduces the formation of In_2Te_3 but, in addition, causes the crystalline structure to deteriorate, leading finally to polycrystalline growth.

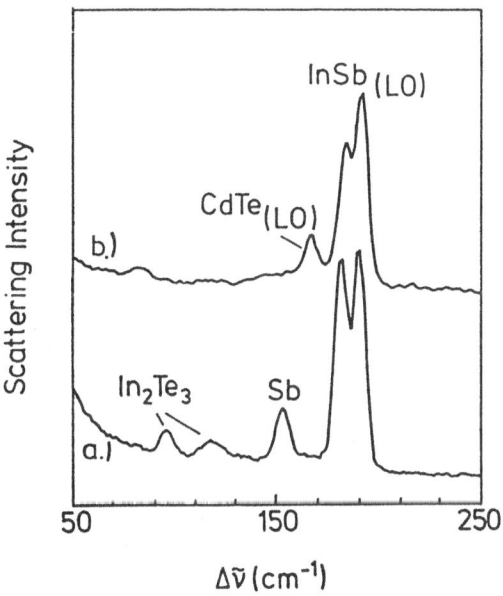

Fig. 5.20. RS of CdTe/InSb(001) heterostructures grown by MBE at 570K: a) with a single CdTe source, b) with Cd overpressure (Cd/Te = 3) [49].

CdS on InP(110), which is a lattice matched system similar to CdTe/InSb, was the first one to be investigated *in situ* [50]. Fig. 5.21 shows the development of the Raman spectra with increasing thickness. The 1 LO- and 2 LO-phonon features of cubic CdS at 300 and 600 cm^{-1}, respectively, are clearly seen. The broad structure in the spectrum of 2 ML appearing below 300 cm^{-1} turns out to be correlated with the formation of In_2S_3 at the interface. At larger thickness this structure is no longer seen because the strong CdS phonon scattering intensities increase with thickness. After deposition of approximately 100 ML the scattering intensities decrease suddenly again due to the fact that the critical thickness has been reached and the overlayer is now relaxing.

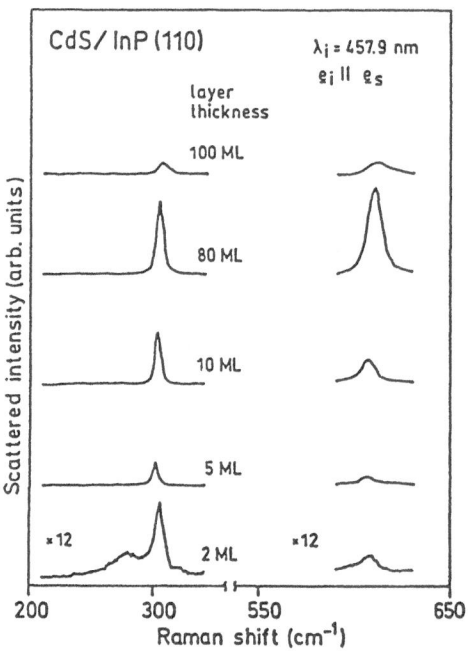

Fig. 5.21. Raman spectra of CdS grown on InP(110) as a function of coverage [50].

The growth of ZnSe on GaAs(110) is followed in Fig. 5.22. At low nominal coverage, a structure at 150 cm^{-1} appears which is due to the formation of Ga$_2$Se$_3$. Simultaneously, the TO and LO phonons of GaAs appear which are forbidden in this polarisation configuration on a (001) surface. This is explained by surface roughening due to Ga$_2$Se$_3$ formation. The first ZnSe signal appears at a nominal 47 nm coverage, and grows with increasing coverage. The conclusion is that, under these conditions, Zn is not adsorbed on the GaAs(001) surface. The epitaxial growth of ZnSe seems to be enabled only after the surface is modified through reactive formation of Ga$_2$Se$_3$..

The knowledge of such exchange reactions at interfaces has also lead to the direct growth of III$_2$VI$_3$ layers by exposing III-V semiconductor surfaces to a flux of elemental or molecular group VI species [52].

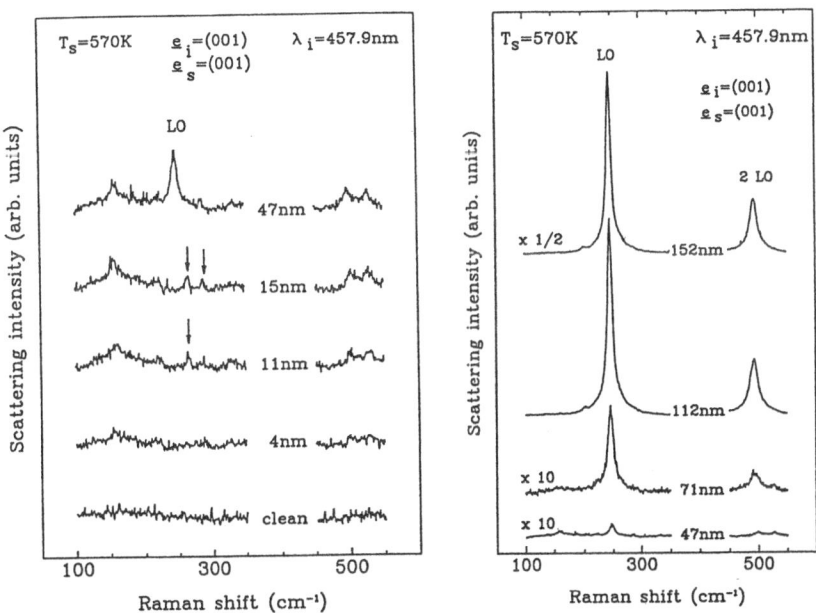

Fig. 5.22. Raman spectra of ZnSe grown on GaAs(110), as a function of the nominal coverage [51].

References

1. Smekal, A.: Naturwissenschaften *11*, 873 (1923)
2. Raman, C.V.: Indian Journal of Physics *2*, 387 (1928)
3. Klein, M.V. in: Cardona, M., Güntherodt, G. (eds.) Topics in Applied Physics, vol. 8; Light Scattering in Solids IV. Springer, Berlin 1975, p. 148
4. Abstreiter, G. in: Cardona, M., Güntherodt, G. (eds.) Topics in Applied Physics, vol. 54; Light Scattering in Solids IV. Springer, Berlin 1975, p. 5
5. Hayes, W., .Loudon, R.: Light Scattering in Solids. Wiley, New York 1978
6. Richter, W. in: Höhler, G. (ed.) Springer Tracts in Modern Physics, vol. 78; Resonant Raman Scattering in Semiconductors. Springer, Berlin 1976, p. 137
7. Cardona, M. in: Cardona, M., Güntherodt, G. (eds.) Topics in Applied Physics, vol. 50; Light Scattering in Solids II. Springer, Berlin 1982, p. 19
8. Richter, W., Nowak, U., Stahl, A.: Proceedings of the International Conference on the Physics of Semiconductors, Kyoto 1980, p. 703 (1980)
9. Born, M., Huang, K.: Dynamical Theory of Crystal Lattices. Clarendon Press, Oxford 1966

10. Pinczuk, A., Burstein, E. in: Cardona, M., Güntherodt, G. (eds.) Topics in Applied Physics, vol. 8; Light Scattering in Solids IV. Springer, Berlin 1975, p. 23
11. Kane, E.O.: Physical Review *178*, 1368 (1969)
12. Fröhlich, H.: Advances in Physics *3*, 325 (1954)
13. Geurts, J.: Surface Science Reports *18*, 1 (1993)
14. Heine, V.: Group Theory in Quantum Mechanics. Pergamon Press, Oxford 1960
15. Loudon, R. : Advances in Physics *13*, 423 (1964)
16. Claus, R., Merten, L., Brandmueller, J. in: Springer Tracts in Modern Physics, vol. 75. Höhler, G. (ed.) Springer, Berlin 1976
17. Hathaway, C.E. in: The Raman Effect, Volume 1. Anderson, A. (ed.) Marcel Dekker Inc., New York 1971
18. Wagner, V., Drews, D., Esser, N., Zahn, D.R.T., Geurts, J., Richter,W.: Journal of Applied Physics *75*, 7330 (1994)
19. Duke, C.B., Patron,A., Ford,W.K., Kahn, A., Carelli, J.: Physical Review *B 26*, 803 (1982)
20. Duke, C.B., Mailhot, C., Patron, A., Li, K., Bonapace, C., Kahn, A.: Surface Science *163*, 391 (1985)
21. Hansson, G.V.: Physica Scripta *T 17*, 70 (1987)
22. Martenson, P., Hansson, G.V., Laedeniemi, M., Magnusson, K.O., Wiklund, S., Nichols, J.M.: Physical Review *B 33*, 7399 (1986)
23. Tulke, A., Mattern-Klosson, M., Lüth, H.: Solid State Communications *59*, 303 (1986)
24. McGovern, I.T., Whittle, R., Zahn, D.R.T., Müller, C., Nowak, C., Cafolla, A., Braun, W.: Journal of Physics: Condensed Matter *3*, S367 (1991)
25. Drube, W., Himpsel, F.J.: Physical Review *B 37*, 855 (1988)
26. Feenstra, R.M., Martenson, P.: Physical Review Letters *61*, 447 (1988)
27. Shih, C.K., Feenstra, R.M., Martenson, P.: Journal of Vacuum Science and Technology *A 8*, 3379 (1990)
28. Harten, U., Toennies, J.P.: Europhysics Letters *4*, 833 (1987). Santini, P., Miglio, L., Benedek, G., Harten, U., Ruggerone, P., Toennies, J.P.: Physical Review *B 42*, 11942 (1990)
29. Hünermann, M., Geurts, J., Richter, W.: Physical Review Letters *66*, 640 (1991)
30. Godin, T.J., LaFemina, J.P., Duke, C.B.: Journal of Vacuum Science and Technology *B 9*, 2282 (1991)
31. Schmidt, W.G., Srivastava, G.P.: Proceedings of ECOSS-14, Surface Science, in press (1995)
32. Richter, W., Esser, N., Kelnberger, A., Köpp, M.: Solid State Communications *84*, 165 (1992)
33. Mailhot, C., Duke, C.B., Chadi, D.J.: Journal of Vacuum Science and Technology *A 3*, 915 (1985)
34. Mailhot, C., Duke, C.B., Chadi, D.J.: Physical Review *B 31*, 2213 (1985)
35. LaFemina, J.P., Duke, C.B., Mailhot, C.: Journal of Vacuum Science and Technology *B 8*, 888 (1990)
36. Sinyukov, M., Trommer, R., Cardona, M.: Physica Status Solidi *(b) 86*, 563 (1978)
37. Esser, N., Köpp, M., Haier, P., Richter, W.: Journal of Vacuum Science and Technology *B 11*,1481 (1993)
38. Huenermann, M., Pletschen, W., Resch, U., Rettweiler, U., Richter, W., Geurts, J., Lautenschlager, P.: Surface Science *189/190*, 322 (1987)
39. Zitter, R.N. in: Carter, E.L., Bate, R.T. (eds.) The Physics of Semimetals and Narrow-Gap Semiconductors. Pergamon Press, Oxford1971, p. 285
40. Richter, W., Fjeldly, T., Renucci, J., Cardona, M. in: Balkanski, M. (ed.) Lattice Dynamics. Flammarion Sciences, Paris 1978, p. 104
41. Resch,U., Esser, N., Richter, W.: Surface Science *251/252*, 621 (1991)

42. Pletschen, W., Esser, N., Münder, H., Zahn, D., Geurts, J., Richter,W.: Surface Science *178*, 140 (1986)
43. Esser, N., Hünermann, M., Resch, U., Spaltmann, D., Geurts, J., Zahn, D., Richter, W., Williams, R.H.: Applied Surface Science *41/42*, 169 (1989)
44. Brugger, W., Schaeffler, F., Abstreiter, G.: Physical Review Letters *52,* 141 (1984)
45. Uhrberg, R.I.G., Bringans, R.D., Olmstead, M.A., Bachrach, R.Z., Northrup, J.E.: Physical Review *B 35*, 3945 (1987)
46. Esser, N., Köpp, M., Haier, P., Richter, W.: Journal of Electron Spectroscopy and Related Phenomena *64/65*, 85 (1993)
47. Wilhelm, H., Richter, W., Rossow, U., Zahn, D., Westwood, D.I., Woolf, D.A., Williams, R.H.: Surface Science *251/252*, 556 (1991)
48. Wagner, U., Drews, D., Esser, N., Richter, W., Zahn, D.R.T., Geurts, J. in: Lengeler, B., Lüth, H., Mönch, W., Pollmann, J. (eds.) Proceedings of the Fourth International Conference on the Formation of Semiconductor Interfaces. World Scientific Press, Singapore 1994, p. 506
49. Zahn, D.R.T., Williams, R.H., Golding, T.D., Dinan, J.H., Mackey, K.J., Geurts, J., Richter, W.: Applied Physics Letters *53*, 2409 (1988)
50. Zahn, D.R.T., Maierhofer, C., Winter, A., Reckzügel, M., Srama, R., Thomas, A., Horn, K., Richter, W.: Journal of Vacuum Science and Technology *B 9*, 2206 (1991)
51. Nowak, C., Zahn, D.R.T., Rossow, U., Richter, W.: Journal of Vacuum Science and Technology *B 10*, 2066 (1992)
52. Zahn, D.R.T., Krost, A., Kolodziejczyk, M., Filz, T., Richter, W.: Journal of Vacuum Science and Technology *B10*, 2077 (1992)

Chapter 6.
Photoluminescence Spectroscopy

Zbig Sobiesierski

Department of Physics and Astronomy, University of Wales College of Cardiff

6.1 Introduction

It is only fitting that a chapter on photoluminescence (PL) should be included in a book which ascribes to reflect the present role of optical techniques in the characterisation of surfaces and interfaces. Whilst the phenomenon of luminescence has been investigated for more than 100 years, it is only within the last 10 to 15 years that luminescence spectroscopies have been applied to semiconductor surfaces and interfaces. This late maturing, of a well-established experimental technique, is linked almost exclusively to improvements in material quality, which have resulted from the ability to deposit epitaxial layers with atomic precision. In particular, the confinement of carriers in a potential well, afforded by growing a thin layer of a material with a bandgap less than that of the surrounding material, means that luminescence measurements become probes of the local electronic environment. Moreover, information can be gathered on specific surfaces and interfaces by monitoring the coupling which occurs between the quantised well levels and the electronic states at the surface or interface itself.

The aim of this chapter is to highlight specific ways in which PL measurements have increased the physical understanding of processes occurring at semiconductor surfaces and interfaces. The discussion concentrates on planar surfaces and interfaces, since these offer the simplest systems which can be imaged with the widest range of experimental techniques.

6.2 History

Luminescence corresponds to the emission of EM radiation, from a body, which occurs subsequent to some excitation process. *Photoluminescence* is a general term used to describe the effect observed following excitation with light (photons). The time interval between excitation and emission, in turn, determines whether the process is fluorescence (almost instantaneous) or phosphorescence (delayed *after glow*). In fact, the emission characteristic of a particular body often contains a wide range of radiative lifetimes. Lecoq de Boisboudran was the first to show, in 1886, that the pronounced luminescence of sulphide phosphors was due to the presence of a foreign element at low concentration. The first systematic study of luminescent crystals was carried out by Lenard and his co-workers [1], followed by the work of Hilsch [2], Smakula [3], Hilsch and Pohl [4]. Lenard had originally considered that the foreign elements, which activated the luminescence, formed an agglomerate with selected atoms of the solid, linked together by valence forces in the manner of an organic ring compound. Such centres thus furnished a mechanism for both excitation and emission.

However, the validity of assigning the luminescence to the presence of colloidal ingredients within the crystal was subsequently questioned by the studies of Schleede and Gantzchov [5]. X-ray patterns, obtained from the sulphide phosphors of calcium, barium, magnesium and zinc, revealed that these luminescent crystals possessed a "normal" crystal structure. In fact these scientists had, between them, already exposed the principle features of extrinsic luminescence in crystals. In addition to the active centres, which needed to be added during crystal growth as a dopant to bring about luminescence, they were also aware of other impurities, whose presence as "killer" centres inhibited the luminescence process.

Figure 6.1 depicts a typical energy versus configuration coordinate diagram, for the process of excitation and subsequent luminescence [6]. In accordance with the Franck-Condon principle [7], light absorption and emission appear as vertical transitions, followed by atomic relaxation of the excited and ground states, respectively. In the case of a crystalline semiconductor, both the initial and excited state levels have a much narrower energy bandwidth than that depicted schematically in the figure.

By the 1960's, the alkali halides were still proving to be a fruitful area for luminescence research. Advances in crystal growth techniques, coupled with the possibility of performing measurements at liquid helium temperatures, meant that the physics of a whole host of colour centres could be investigated. Furthermore, the ability to perform temperature dependent measurements, from 4.2K to above room temperature, allowed the assessment of the contribution from the lattice vibrations to the luminescence lineshape [7]. This meant that the strength of the electron-phonon interaction could be assessed in a wide variety of materials. The symmetry of luminescence arising from strongly coupled defects provided a

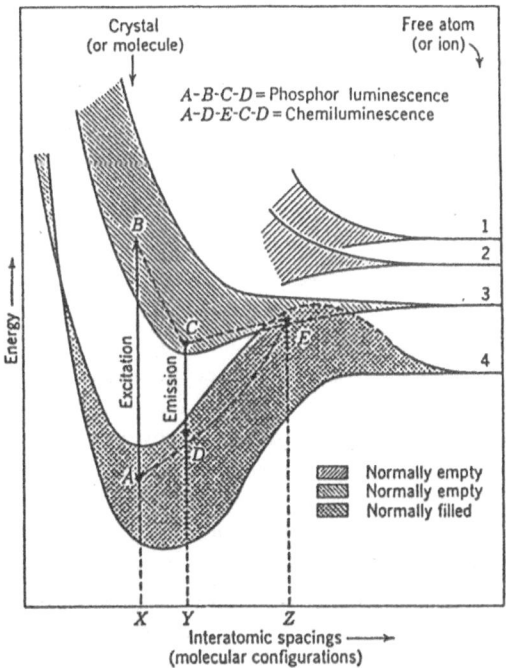

Fig. 6.1. Generalised configuration-coordinate type of energy level diagram for a hypothetical phosphor or luminescent molecule [6].

measure of the crystal structure surrounding the defects themselves, as well as the localised phonon modes. At the same time, the *exciton* (see, for example, Ref. 8), an entity which had already been conceived theoretically in the 1930's, was being researched vigorously.

The subsequent improvement in purity of source materials meant that layers could be produced with a greatly reduced number of extrinsic defects, thereby allowing the intrinsic recombination processes to be examined in bulk semiconductor crystals. The development of advanced crystal growth techniques, such as both MBE and vapour phase epitaxy (VPE), meant that layered structures of clearly defined composition could be grown on an atomic, layer-by-layer basis. In fact, it is precisely the study of the radiative recombination and dynamics of excitons, in addition to that of electrons and holes, together with the ability to fabricate epitaxial structures on an atomic scale, which has led to the increasing use of luminescence measurements in the characterisation of low-dimensional semiconductors.

6.3 Experiment

The main components of a PL system are laid out, in modular fashion, in Fig. 6.2. With this approach, signal extraction can occur either by phase-sensitive detection, or photon counting techniques. In phase-sensitive detection, where the excitation is modulated, the luminescence signal corresponds to the coherent component of the signal, measured at the frequency of modulation. On the other hand, the photon counting approach integrates all the recorded events which have an amplitude greater than that of the background noise, as determined by the setting of a discriminator level. The source of excitation has evolved from a discharge lamp, filled with one or more gases, through to the wide range of both continuous and pulsed lasers available today. The ability to produce pulses of light, which last for only a fraction of a picosecond, has provided the perfect tool to probe the dynamics of photoexcited carriers on a time scale of the competing non-radiative processes. At the same time, the development of dye lasers and, more recently, lasers based on doped crystals such as Ti-sapphire, has allowed the generation of narrow bandwidth excitation which can be tuned continuously in energy. This latter development has directly led to an upsurge in both the use and application, of *photoluminescence excitation spectroscopy* (PLE), a technique that allows the identification of the excited states which contribute to a particular luminescence transition. The sample under investigation is generally mounted in some form of cryostat, which usually allows the sample temperature to be varied continuously from 4.2K to 300K.

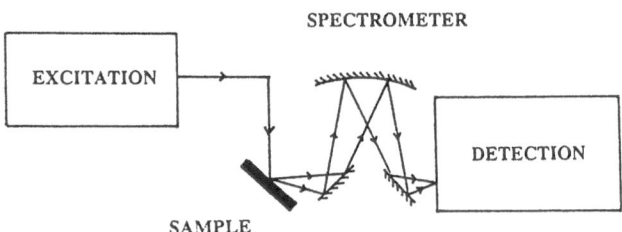

Fig. 6.2. Principal components of a dispersive photoluminescence measurement system.

Light emitted from the sample is collected, using an appropriate optical arrangement, and coupled into some form of spectrometer which contains a dispersive element, such as a prism or diffraction grating. The function of the spectrometer is to provide an energy spectrum of the collected light, with the highest spectral purity (best S/N ratio) across the plane of the exit slit. To this extent, the development of the diffraction grating (ruled or holographic) has provided a dispersive element with both a greater degree of dispersion and a higher light throughput, than any equivalent prism. The experimental arrangement required for PLE remains identical to that depicted schematically in Fig. 6.2. However, in this case, it is the energy of excitation which is scanned whilst the spectrometer is set to detect a particular PL transition energy.

More than 50 years ago, the most common ways of detecting luminescence relied on either optogalvanic, or photoconductive cells such as PbS and PbSe. The invention of the photomultiplier tube, possessing a much greater sensitivity, meant that weak luminescence signals could be measured in the visible and UV range of the spectrum. A degree of response was available in the near-IR, at wavelengths out to 1100 nm, with the S 1 photocathode. However, it was not until the advent of devices based on high-purity layers of Si and Ge that luminescence could be detected in the mid-IR region up to 1700 nm. In fact, Ge detectors, in which both the Ge p-i-n structure and the preamplifier are liquid nitrogen cooled, still provide the best way of acquiring a complete luminescence spectrum in the wavelength range 700 to 1700 nm. More recently, the controlled processing of Si wafers has resulted in the fabrication of two dimensional arrays of discrete diodes. The existence of diode arrays, or charge-coupled devices (CCDs), has meant that entire optical spectra can be acquired simultaneously, thus significantly speeding up the time required to perform a measurement. In addition, the capacity to take snap-shots of the electronic system under investigation, whilst subjected to an external perturbation, has increased the precision of the measurement in question. For example, in pressure-dependent studies of the luminescence behaviour in semiconductors, the luminescence from a piece of ruby is normally used to calibrate the actual pressure exerted on the sample. The use of a CCD allows both spectra to be acquired simultaneously, providing the luminescence energies are not too dissimilar.

The other fast way of acquiring complete spectra is to use a Fourier transform spectrometer, as illustrated in Fig. 6.3. Light collected from the source is split into two equal beams by the beamsplitter B, each of which is reflected off a mirror M1 or M2 before being recombined at B to form a single beam, which is subsequently detected. The degree of interference between the two beam paths, path difference 2(BM1 - BM2), can be varied by altering the position of the moveable mirror M1. If, for example, the light source only contains a single frequency, then a sinusoidal waveform will be detected if the path difference is changed in a continuous manner. Hence, the measured signal is the Fourier transform (in the time domain) of the spectrum of the source (in the frequency domain). This means that the time required to obtain a complete PL spectrum is reduced to the time needed to acquire the Fourier transform in the time domain and perform the reverse transform. With the advent of the fast Fourier transform algorithm [9], the total time required is of the order of a few seconds. At this point, the reader may well be wondering why Fourier spectroscopy is not used to acquire all spectral data. The chief drawback lies in the restricted dynamic range of the method: weak signals cannot be resolved in the presence of much more intense features. Hence, as far as PL measurements are concerned, dispersive spectrometers together with either single channel or parallel detection, remain the most versatile tools for spectroscopists.

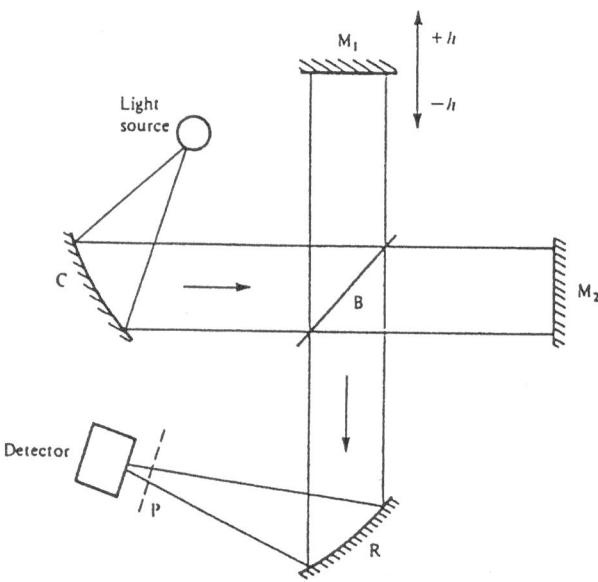

Fig. 6.3. Twyman-Green version of a Michelson-type Fourier-transform spectrometer.

6.4 Basic Principles

As mentioned in Sect. 6.2, PL corresponds to the emission of EM radiation following photoexcitation of a body. In the discussion which follows, the system under investigation will be confined to that of interband transitions in semiconductors. The threshold energy required for photoexcitation, in this case, corresponds to the energy needed to transfer an electron from a bonding (valence) state to a non-bonding (conduction) state, and is thus equivalent to the energy required to break a bond (see, for example, Refs. 10, 11). Figure 6.4 depicts the photoexcitation process in terms of both real space and energy levels. The transfer of an electron out of the valence band leaves behind an unoccupied level, or hole, which is positively charged. The PL process is then completed by the radiative recombination, and thus annihilation, of an electron-hole pair.

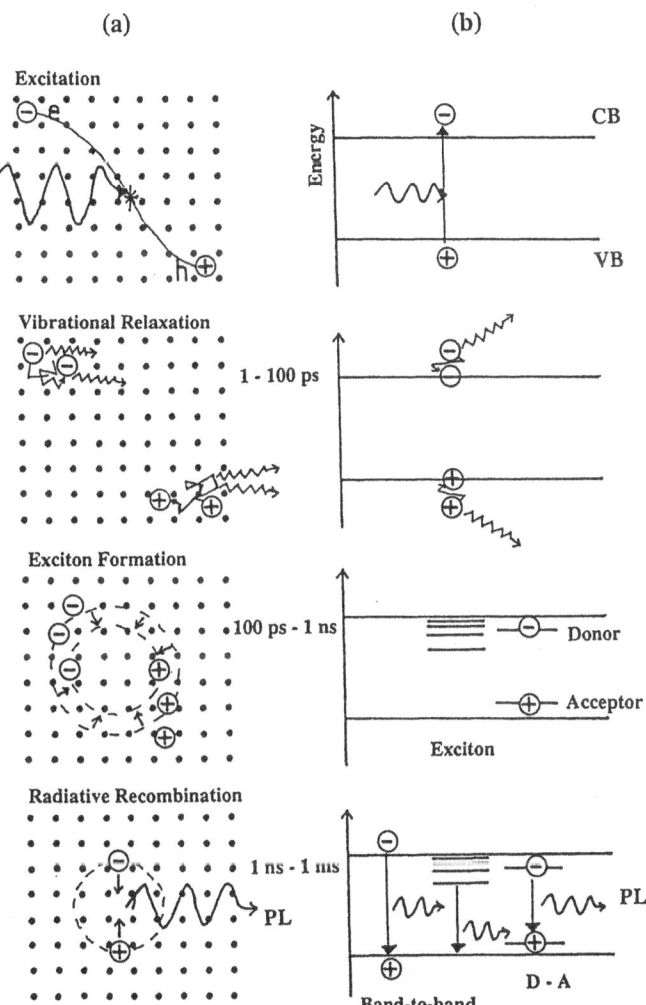

Fig. 6.4. Visualisation of photoexcitation, carrier relaxation, and subsequent radiative recombination, on both (a) real space and (b) energy level diagrams.

There are a number of separate recombination processes which can occur, as shown in Fig. 6.4b. These correspond to transitions involving free carriers, free carriers to donor or acceptor levels; transitions between carriers trapped by donor and acceptor states, and those involving both free and bound excitons. Since donor-acceptor recombination involves carriers with localised wavefunctions, it can only take place directly by radiative tunnelling between the two energy levels.

As such, the PL energy will contain a Coulomb term, which reflects the physical separation, r, of the two trap sites in real space:

$$\hbar\omega = E_g - (E_A + E_D) - (e^2/4\pi\varepsilon r) \tag{6.1}$$

where $\hbar\omega$ is the energy of the emitted photon, E_g the bandgap of the semiconductor, E_A and E_D the binding energies of acceptors and donors, respectively, and ε is the dielectric function of the material. In most cases, the wide distribution of donor-acceptor separations merely results in a broadening to the width of the PL band. However, in a few cases [12-14], the presence of well-defined donor and acceptor sites results in a PL spectrum containing a great deal of fine structure, and a wealth of information which can be deduced by fitting an appropriate donor-acceptor distribution function to the PL data. Figure 6.5 provides an illustration, with PL spectra obtained from GaP crystals, after selective doping with either Zn or Cd atoms. The multitude of fine lines reflects a wide range of donor(S)-to-acceptor(Zn or Cd) pair separations.

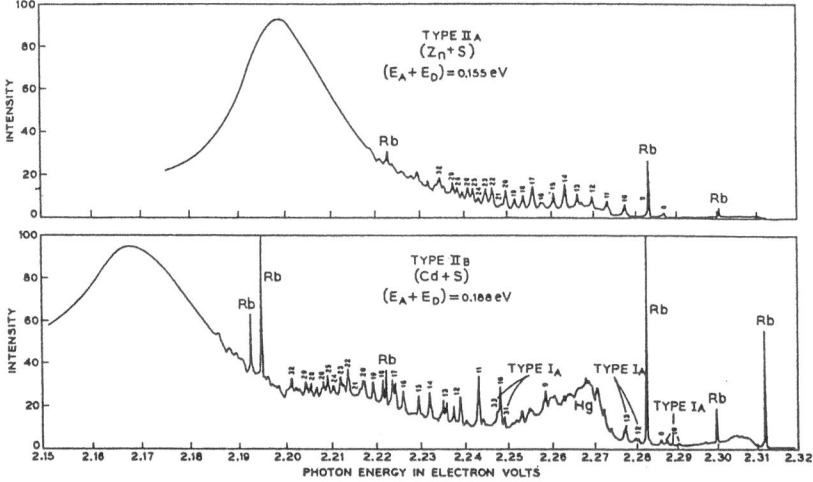

Fig. 6.5. 1.6K PL spectra obtained from GaP crystals, to which either Zn or Cd atoms have been added. The sharp lines correspond to a wide range of distinct donor-acceptor pair separations [12].

An exciton is formed when a photoexcited electron-hole pair are brought together by means of their mutual Coulomb interaction. Since an exciton is electrically neutral, it can often diffuse over much greater distances (at low temperature) than isolated charge carriers, without being either scattered or recombining. Moreover, the correlation of an electron with a specific hole, means that an exciton can be described analogously to a hydrogen atom, in terms of an electron and hole orbiting about a reduced centre of mass. As such, there exists a certain binding energy which needs to be supplied in order to ionise the exciton, together with a characteristic physical extent to the exciton, defined in terms of an effective Bohr radius (typically 4-40 nm). The excitonic radius does not play much of a role in bulk material; however, as will be discussed later, it does enable PL measurements to become extremely sensitive probes of semiconductor interfaces, when the dimensions of the layer become comparable with that of the exciton.

In addition to the peak energy of a particular PL transition, the intensity and full-width-half-maximum (FWHM) values of the lineshape are of interest. Both parameters are affected by the quality of the crystal, in that a greater number of impurities, or defects, tends to both decrease the intensity, and increase the FWHM of the PL peak. The FWHM, in particular, reflects the lifetime of the carriers involved in the recombination process. It is inversely proportional to an effective lifetime, τ_{eff}, which is given by:

$$\frac{1}{\tau_{eff}} = \frac{1}{\tau_r} + \frac{1}{\tau_{nr}} + \frac{1}{\tau_s} \tag{6.2}$$

where τ_r is the radiative lifetime, τ_{nr} the non-radiative lifetime arising from the bulk, and τ_s the non-radiative lifetime arising from the surface or interface. Hence, by appropriate selection of the sample structure, it is possible to focus on either bulk, or surface/interface recombination, independently of each other. For example, in the case of low- to moderately-doped thick epitaxial layers, the electrical mobility will be inversely proportional to the FWHM of the PL signal.

Similarly, interfacial recombination at a particular heterojunction can be explored in detail if the sample is terminated with a wider bandgap material, which then acts as a mirror to prevent both free electrons and holes from reaching the surface. If the region of interest is now sandwiched between layers of wider bandgap material, it will form a potential well for either, or both, electrons and holes. As the width of the well region is reduced, a point is reached at which the motion of carriers becomes restricted to the plane of the well. In this manner, the degeneracy of the states has been partially lifted, and the population of the well can be described in terms of electron and hole gases with defined states of energy quantisation. In the simplest case of a well with infinite potential walls, the allowed energy levels are given by:

$$E_n = (\pi^2 \hbar^2 / 2)(n^2 / m^* L^2) \tag{6.3}$$

where m^* is the effective mass of the electron (hole), L is the width of the quantum well (QW), and n is a quantum number describing the state of excitation. The refinement associated with computing the energy levels within a well with finite barriers will not be dealt with here: the interested reader is referred to the excellent review article by Herman *et al* [15] for additional information. Since the thickness at which a potential well begins to show quantum effects is similar to the size of the exciton radius, it follows that the shape of the exciton will become distorted or squeezed as the thickness of the QW is reduced. This physical extension of the exciton, in a direction parallel to the well interfaces, means that the excitonic luminescence will reflect the interfacial roughness.

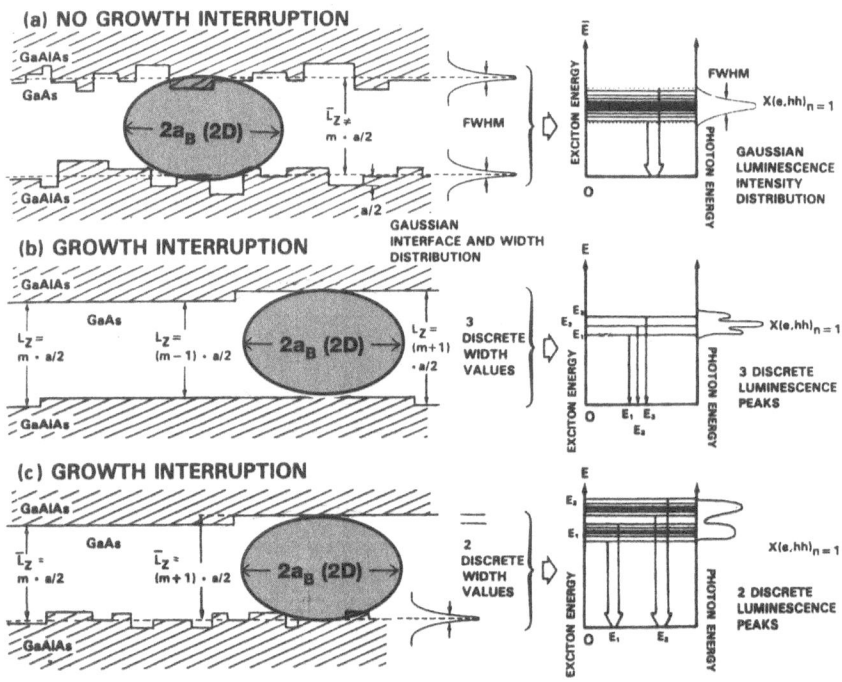

Fig. 6.6. Model for the interface disorder at GaAs/AlGaAs quantum well interfaces, together with predicted effect on PL lineshape arising from excitonic recombination [15].

Figure 6.6 provides three examples of the kinds of PL spectra which might be expected for interfacial roughness containing different length scales [15]. With the existence of large, monolayer-flat terraces at both interfaces, discrete components in the PL spectrum can be related directly to monolayer fluctuations in thickness at the well interfaces. On the other hand, the absence of large terraces at either interface leads to a broadening of the PL lineshape. However, the presence of microroughness, at either or both interfaces, can also result in sharp luminescence features [15, 16]. In this case, the energy splitting between PL peaks does not relate directly to fluctuations in thickness, on a monolayer scale, and may vary between different positions on the sample.

At this point, it is worth emphasising the bearing which the choice of material system has on the interpretation of PL data from low-dimensional structures. As noted earlier, the FWHM of the PL lineshape in bulk materials scales with the number of scattering centres present. The addition of one or more interfaces requires that the interfacial roughness also be included in any calculation of the line broadening [17, 18]. Hence, as the thickness of a QW is reduced, one would expect intuitively the FWHM of the PL spectrum to increase as a result of the increased surface-to-volume ratio. In the case of GaAs/AlGaAs, this is precisely what is observed. However, PL spectra obtained from narrow wells fabricated in the InGaAs/GaAs system exhibit FWHMs which decrease with decreasing well thickness and, moreover, have values below those expected for alloy scattering in bulk InGaAs layers.

This difference in behaviour can be accounted for by considering the penetration of the electron and hole wavefunctions into the barrier regions, on either side of the potential well. For wide wells, with semi-infinite potential barriers, the wavefunctions are confined solely to the well region. However, with decreasing well width, the wavefunctions penetrate further into the barrier layers and, correspondingly, sample the electronic environment within these regions. For GaAs/AlGaAs structures, the AlGaAs barrier layers have greater alloy scattering than the GaAs well region, hence the PL linewidth will continue to increase with decreasing well width. However, for the InGaAs/GaAs system, the GaAs barrier layers have a lower value for alloy scattering than the InGaAs well material, which means that the PL linewidth begins to decrease below a certain value of well width.

A final point relates to the sensitivity of PL measurements as a probe of interfaces in low-dimensional structures. The luminescence corresponds to the recombination of all carriers photoexcited within a diffusion length of the well region and, subsequently, trapped in the well. This means that the signal is much stronger than that expected for recombination of carriers excited within the well region alone by direct absorption. Hence, at low temperatures, where thermal excitation of the carriers out of the well region becomes negligible, resolvable PL emission can still be observed even for ultra-thin (1-2 ML thick) OWs.

6.5 Selected Examples

6.5.1 Recombination at Semiconductor Surfaces and Interfaces

The surface recombination velocity, S, can be defined as the number of carriers which recombine at a surface or interface per unit area per unit time, per unit volume of excess bulk carriers at the boundary between the quasi-neutral and space-charge regions. This is schematically depicted by Fig. 6.7a, in which W is the effective width of the space charge region. As outlined by Aspnes [19], S can range up to an appreciable fraction of the thermal speed of free carriers. The critical range for device purposes is $S \sim 10^3$ to $\sim 10^7$ cm s^{-1}, as discussed by Ellis and Moss [20]. Surface recombination becomes less important than volume recombination for values of S substantially below 10^3 cm s^{-1}.

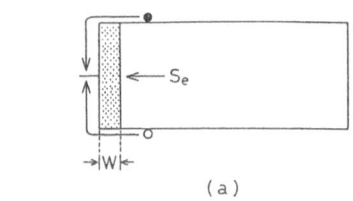

(a)

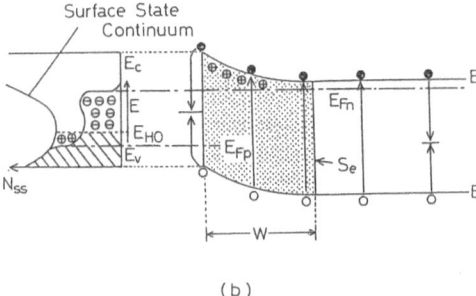

(b)

Fig. 6.7. (a) Simple model for surface recombination, with constant values for surface recombination velocity, S, and depletion layer, W. (b) A more realistic model, in which both S and W depend on the actual excitation density used, together with an appropriate surface state distribution [31].

PL has been used extensively as a measure of the changes in S, which can be effected by passivation treatments applied to semiconductor surfaces. For example, a value for S of 450 cm s^{-1} has been obtained at the Ga$_{1-x}$Al$_x$As/GaAs interface [21, 22], three orders of magnitude smaller than at the air/GaAs interface. However, the restriction of having to deposit AlGaAs, added to the degradation of AlGaAs layers with time, has resulted in considerable effort being applied to the chemical passivation of GaAs surfaces (23-26). Similar attention has been focused on passivation of InP, with PL measurements also being performed *in situ* during various surface treatments [27]. Whilst effective passivation has clearly been demonstrated, the long term stability remains an unresolved problem [28].

Conventionally, integrated PL intensity and PL intensity decay curves have been used to produce the appropriate Shockley-Read-Hall [29, 30] expression for the surface recombination. This has the limitation of assuming constant values for S and W, irrespective of the excitation conditions under which the data were acquired. More recently, Saitoh *et al* [31] have refined the calculation to include the dependence of S and W on the actual excitation intensity used. In this case, a one-dimensional Scharfetter-Gummmel [32] simulation program has been used to establish the required self-consistency between surface and bulk recombination, carrier flows, charge distributions and interface state occupation, as shown by Fig. 6.7b. Figure 6.8 displays the measured PL efficiency curves, together with the calculated theoretical plots, for GaAs surfaces subjected to a number of different treatments [31]. The inset shows the distribution of surface states which provided the best fit to the experimental data. Apart from the differences predicted in interface state densities, N_{ss} and N_{ss0}, the sulphur-treated case can only be explained by assuming generation of a significant amount of negative fixed charge, Q_{fc}. Hence care needs to be taken when attributing changes in band bending to passivation effects, since band bending is caused by the *total* surface charge, which is the sum of charge related to interface states in the gap and fixed charge on the surface, as stressed by Yablonovitch *et al* [25]. Figure 6.9 contains a similar comparison between experiment and theory for InGaAs surfaces passivated with SiO_2, both with and without a thin Si interfacial layer.

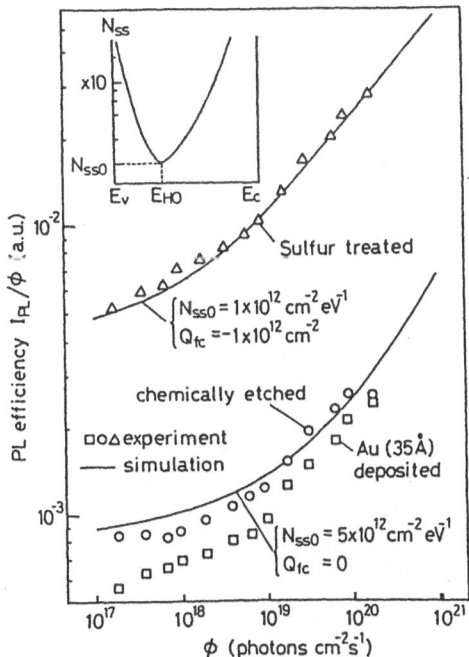

Fig. 6.8. PL efficiency spectra for GaAs surfaces subjected to various treatments [31].

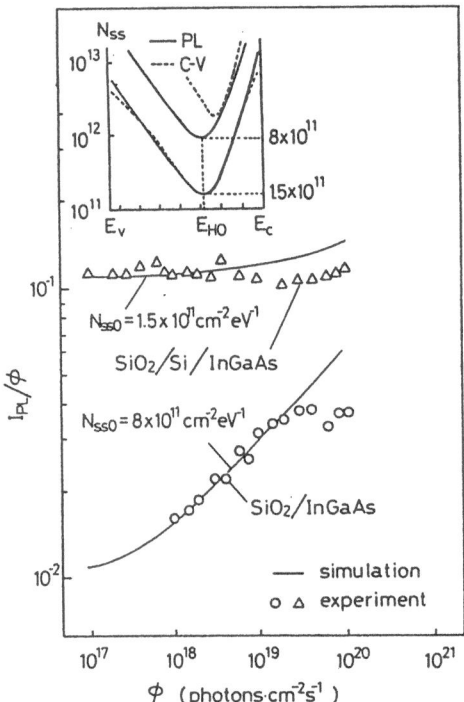

Fig. 6.9. PL efficiency spectra for passivated InGaAs surfaces, with and without a Si interface-control-layer [31].

Thus far, only changes in integrated luminescence intensities have been discussed. Spectrally-resolved PL measurements, on the other hand, can provide valuable information on the energy distribution of radiative surface states, particularly in situations when competing non-radiative channels have been passivated. PL emission, associated with surface roughness [33, 34] and variations in the near-surface chemical composition [35], have previously been reported for InP crystals. Chemical treatment of CdTe, which influences the subsequent Schottky barrier, ϕ_b, obtained for metals deposited on the etched surfaces, have also been shown to lead to drastic changes in the PL spectra [36]. Furthermore, the energies of the dominant deep levels identified in these PL spectra appear to mirror the pinning behaviour observed in the measurements of ϕ_b. Figure 6.10, from the work of Sobiesierski *et al* [36], displays PL spectra obtained at 4.2K from chemically treated CdTe surfaces. XPS measurements have also been performed on these samples, to determine the near-surface Cd/Te ratio, and Au and Sb diodes fabricated on CdTe surfaces etched in a similar manner. In general, etching solutions which leave the n-CdTe surface slightly deficient in Te produce barrier heights of 0.93 eV, while surfaces which have an excess of Te result in values of

ϕ_b of either 0.72 eV or 0.93 eV [37]. The Br/methanol etched surface shows a PL band at 0.875 eV, which is more intense than the PL peaks at 1.125 eV and 1.4 eV. XPS measurements, for this treatment, reveal an excess of Te with respect to the clean cleaved CdTe surface. When the Br/methanol treatment is followed by either a reducing etch, KOH, or hydrazine, the etched surfaces all become restored to stoichiometry, or are slightly deficient in Te. The corresponding PL spectra show a dramatic change, with the 0.875 eV PL band disappearing completely, and the intensities of the peaks at 1.125 eV and 1.4 eV increasing greatly. If one allows for the CdTe bandgap being ~0.16 eV larger at 4.2K than 300K (the temperature at which ϕ_b is determined), then there appears to be remarkable agreement between $\phi_b = 0.72$ eV and the PL transition at 0.875 eV, as well as $\phi_b = 0.93$ eV and the PL transition at 1.125 eV. Hence these two deep levels observed in the PL spectra, which are related to the composition of the CdTe surface, show a close correlation in energy to the values of ϕ_b obtained from diodes fabricated on CdTe surfaces treated in a similar manner.

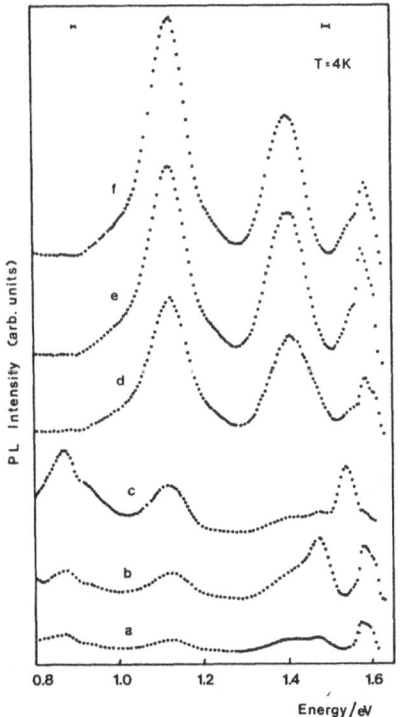

Fig. 6.10. PL spectra for p-CdTe samples chemically treated with (a) an oxidising etch, (b) oxidising + reducing etches, (c) Br/methanol, (d) Br/methanol + reducing etch, (e) Br/methanol + KOH, and (f) Br/methanol + hydrazine.

The remaining PL spectrum in Fig. 6.10 corresponds to the application of an oxidising etch which leaves the near-surface composition with a slight excess of Te. It is worth noting that surface preparations which produce a slight excess of Te (deficiency of Cd) exhibit by far the lowest PL intensities overall. This can be ascribed to increased surface recombination, and an increase in surface leakage currents has also been observed for diodes formed on similar CdTe surfaces [38].

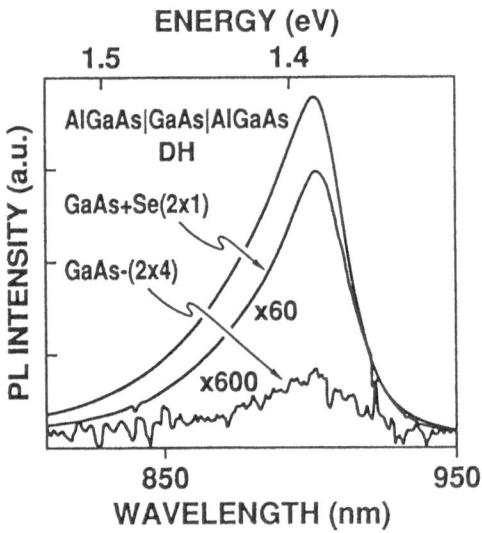

Fig. 6.11. Typical *in situ* PL spectra (at 100°C) from GaAs incorporated in a double heterostructure , compared to PL from a bare (2x4) and a Se-passivated (2x1) surface. PL intensities vary by more than a 1000-fold depending on surface termination [39].

More recently, Sandroff and co-workers [39] have reported PL measurements, conducted *in situ*, on GaAs surfaces where the surface reconstructions and chemical terminations are carefully controlled inside an MBE chamber. PL spectra have been used to distinguish between different GaAs surface reconstructions, as well as assessing several different passivation schemes. Figure 6.11 displays PL spectra, measured *in situ* at 100°C, comparing the emission from an AlGaAs/GaAs double heterostructure with the luminescence from both a bare (2x4) GaAs(001) surface and a Se-passivated (2x1) GaAs(001) surface. Whilst the overall shape of the PL spectra remains the same, it can be seen that the intensities vary by over three orders of magnitude. In addition, the same authors have studied the evolution of the recombination properties of AlGaAs/GaAs interfaces following deposition of successive monolayers of AlGaAs. Figure 6.12 depicts how PL intensity measurements have been used to probe the evolution of the electronic properties at an $Al_{0.5}Ga_{0.5}As$/GaAs interface at early stages of growth. The PL intensity of the GaAs improves only after 4 ML of $Al_{0.5}Ga_{0.5}As$ have been deposited. The similarity of the data obtained at 25°C and 100°C indicates that tunnelling through the $Al_{0.5}Ga_{0.5}As$ layer controls the PL intensity for barrier thicknesses in excess of 4 ML.

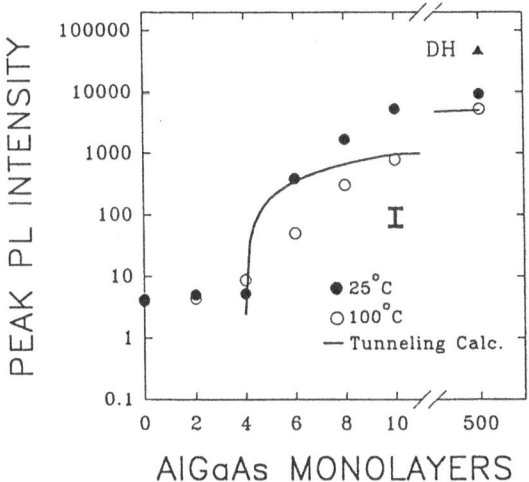

Fig. 6.12. Evolution of electronic properties of the $GaAs/Al_{0.5}Ga_{0.5}As$ interface at early stages of growth. PL from GaAs improves only after 4 ML of $Al_{0.5}Ga_{0.5}As$ have been deposited. The similarity of 25°C and 100°C data indicates that, above 4 ML, tunneling through the $Al_{0.5}Ga_{0.5}As$ layer controls the PL intensity [39].

6.5.2 Surface Quantum Wells

Following the advances made in the understanding and control of radiative recombination occurring within bulk QWs, it has been possible to extend luminescence investigations to surface QWs. Perhaps the most studied, and best characterised, surface QW system is that of InAs/InP(001). This has arisen because clean and well-ordered InP(001) surfaces have conventionally been prepared, for use in growth by MBE, by annealing InP under an As over-pressure. Originally, such thermal cleaning was thought to leave a layer of As on the substrate surface [40]. However, more recent work [41, 42] has shown that oxide removal, under As stabilisation of the InP surface, can result in the exchange of As for P atoms, leading to the formation of a thin pseudomorphic InAs layer on the InP surface. Whilst the surface science experiments of Moison *et al* [41] provided definitive evidence for the epitaxial growth of an InAs surface on bulk InP under these conditions, it was not until the work of Viktorovitch *et al* [43] that PL was observed from similar layers. In fact, it was Cohen and co-workers [44] who suggested that surface QWs might provide a new class of surface light emitters.

Radiative recombination has also been studied in $In_{0.53}Ga_{0.47}As/InP$ surface QWs, prepared by selectively etching away the top InP layer of an $In_{0.53}Ga_{0.47}As/InP$ double heterostructure [45]. The exact thickness and quality of the InAs/InP layers, formed as a result of the As/P exchange reaction, obviously depends on the preparation conditions. However, a combination of XPS and AES experiments [41], together with x-ray photoelectron diffraction [42] and angle-resolved photoemission [46] measurements, indicate a thickness of 1-2 ML InAs. In the work of Sobiesierski *et al* [47], the thickness of the InAs layer on InP(001) has been increased by depositing further epitaxial layers of InAs, using MBE. This has allowed the quantised PL transitions to be measured as a function of InAs well width. Figure 6.13 displays 4.2K PL spectra obtained from 0.25, 0.5, 1.0 and 2.0 nm thick layers of InAs, deposited on InP(001) by MBE. There is a continuous shift of the InAs-related PL emission to lower photon energies, together with an accompanying decrease in PL linewidth, with increasing layer thickness. This observation clearly indicates the formation of InAs QWs on the InP surface. The PL lineshapes, for the InAs layers with deposited thicknesses below 1 nm, resemble those previously attributed to monolayer fluctuations in ultra-thin QW structures [48].

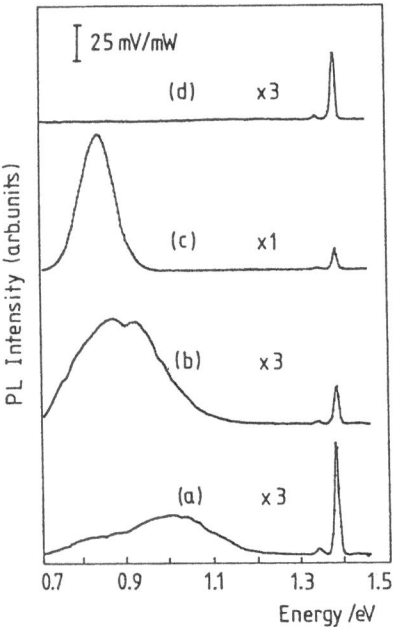

Fig. 6.13. 10K PL spectra, acquired within a week of sample growth, for (a) 0.25 nm InAs/InP, (b) 0.5 nm InAs/InP, (c) 1 nm InAs/InP and (d) 2 nm InAs/InP [51].

In fact, the PL spectra for these layers can be fitted with PL peak energies of 0.83 eV, 0.97 eV and 1.08 eV [47]. Figure 6.14 plots the observed PL peak energy (left-hand axis) together with the integrated InAs-to-InP PL intensity ratio (right-hand axis) versus total InAs thickness (deposited InAs + 0.5 nm InAs due to As/P exchange). The solid curve has been calculated assuming the PL transition to be between the first electron and heavy hole states (e1-hh1) within strained InAs QWs with InP barriers. The value of the conduction band offset, is taken to be $\Delta E_c = 0.4\,\text{eV}$. The agreement between theory and experiment is surprising, particularly as the calculation has: (i) assumed that the potential wells are exactly rectangular; (ii) not included any strain relaxation for the thicker wells; (iii) not corrected for the exciton binding energy. The integrated PL intensity in Fig. 6. 14 is observed to reach a maximum value for a total InAs thickness of 1.5 nm and thereafter decreases with increasing InAs thickness. This decrease in PL intensity is certainly a sign of strain relaxation via the generation of dislocations.

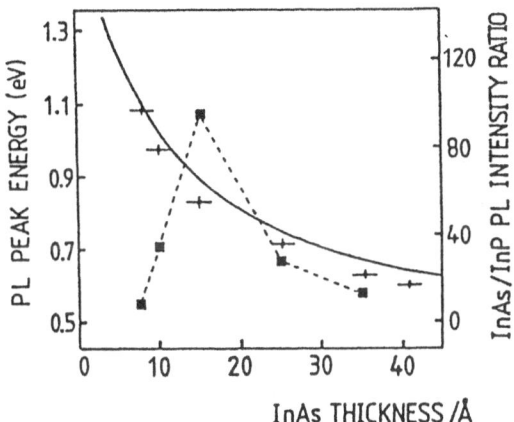

Fig. 6.14. Variation of PL peak energy (+) and InAs-to-InP PL intensity ratio (■) with total InAs layer thickness. Solid line is a calculation for strained InAs/InP quantum wells, taking $\Delta E_c = 0.4\,\text{eV}$ [47].

In situ RHEED measurements of the InAs surface lattice constant have confirmed the onset of relaxation to occur for InAs layers of a similar thicknesses [49]. Evidence of strain relaxation in InAs/InP single QWs, grown by MOVPE [50], has also been identified for well thicknesses in excess of 5 ML InAs. The FWHM values of the PL spectra in Fig. 6.13 reflect both the material quality of the InAs well and the interfacial roughness. The smoothness of the InAs/InP interface is controlled by the As/P exchange reaction, whilst the nature of the InAs/air interface is determined by the oxidised InAs surface. Capping the surface QWs, with 2 ML of GaAs, has been shown to reduce the PL FWHM from ~ 93 meV to ~ 63 meV [47]. The continued oxidation of the exposed InAs surface with time is, in turn, reflected by decreases in the observed PL intensities, coupled with the emergence of higher energy PL components due to effective thinning of the QW [51].

6.5.3 Near-Surface Quantum Wells

Isolated QWs with bulk material barriers form, together with surface QWs, the opposite limits of an entity which is the near-surface QW. By varying the thickness of the barrier, between a QW and the sample surface, it is possible to examine what effect the surface states of the barrier/well material have on the confined electronic states within the well itself. Since the surface-to-volume ratio increases dramatically as the dimensionality of a system is reduced, the influence of surface states will become increasingly significant for quantum wire and quantum box structures. Hence, these studies are of general importance for low-dimensional systems. To simplify the experimental analysis, it is important to choose a system where the number of interface states, generated between the well and barrier, is extremely low. Moison and co-workers [52], have performed *in situ* PL measurements on GaAs/Al$_{0.3}$Ga$_{0.7}$As QWs grown by MBE, where the top Al0.3Ga0.7As barrier thickness has been increased in stages from 0 to 100 nm.

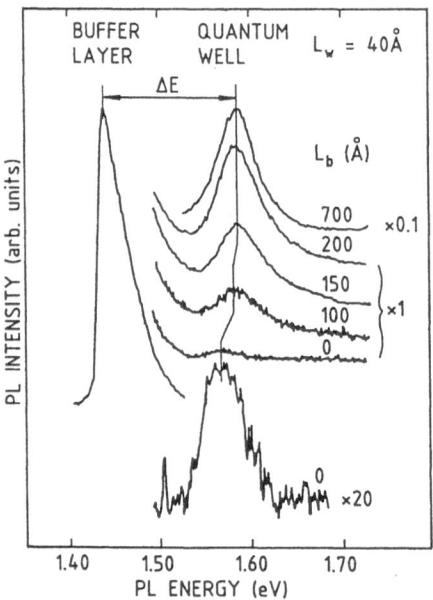

Fig. 6.15. Typical series of PL spectra obtained *in situ* at 180K, for a well width of 4.0 nm and various values of surface barrier thickness, L_b [52].

Whilst this has meant repeated thermal cycling of the sample, from the growth temperature down to the temperature required for PL measurements (180K), a combination of RHEED, LEED, XPS and AES measurements reveal no contamination of the growth surface. This finding has been attributed to the structure of the c-(4x4) surface, whose As adlayer [53, 54] is removed thermally at the beginning of each growth, together with any species adsorbed during the PL analysis, thus leaving a clean surface for epitaxy. Figure 6.15 shows a sequence of PL spectra, which correspond to surface barriers ranging in thickness from 0 to 70 nm AlGaAs. The luminescence arising from the GaAs buffer layer provides an additional measure of the radiative efficiency of the sample. The QW PL both increases in intensity and shifts to higher energy with increasing AlGaAs thickness. Figure 6.16a shows the variation, with barrier thickness, of the

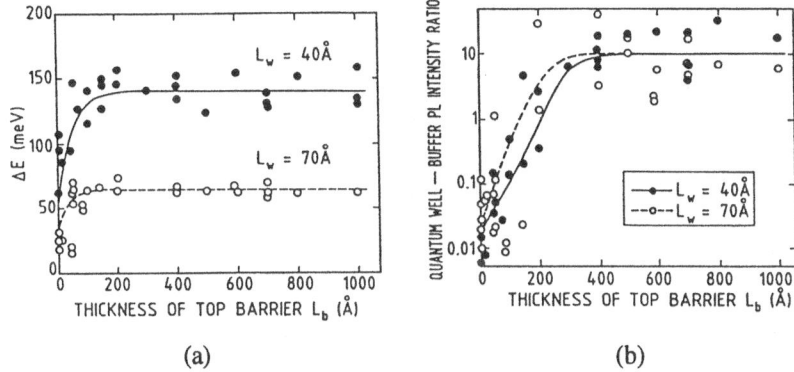

(a)   (b)

Fig. 6.16. (a)Variation of ΔE with L_b for L_w = 4.0 nm (solid circles) and 7.0 nm (open circles), together with fits to the data using the model discussed in the text, and (b) variation of R with L_b for L_w = 4.0 nm (solid circles) and 7.0 nm (open circles), together with a fit to the data, taking $\tau(\infty)/\tau_s$ = 500 [52].

difference in energy between the GaAs buffer layer PL, and the PL signal arising from the near-surface GaAs QW, whilst Fig. 6.16b displays how the intensity ratio of these two PL peaks varies with surface barrier thickness. In both cases, the solid lines represent best fits to a simple model used to describe the electronic coupling between the sample surface and the QW. The interaction between the lowest QW state, $|\psi_w>$, at energy, E_w and a single surface state, $|\psi_s>$, at energy, E_s, leads to a coupled state, $|\psi'>$, at E'. The corresponding wavefunctions, for the surface and QW states, are described by an exponential with a decay length, L_s, for $|\psi_s>$, and a truncated sinusoid for $|\psi_w>$. $|\psi'>$ and E' have then been calculated by first-order perturbation theory, using a square potential value of -0.25 eV. Fitting the observed energy shift in Fig. 6.16a, using E_s and L_s as adjustable parameters, has yielded E_s = 20 ± 10 meV below the bottom of the $Al_{0.3}Ga_{0.7}As$ conduction band (230 meV above the GaAs conduction band edge), and $L_s = 8.0 \pm 1.5$ nm. These values have then been used, together with a further adjustable parameter which reflects the ratio of the bulk-to-surface recombination rates, to fit the PL intensity data in Fig. 6.16b. Moison *et al* [52], have proceeded to discuss the physical relevance of these parameters in terms of the electronic structure of the GaAs(001) and $Al_{0.3}Ga_{0.7}As(001)$ surfaces.

Modification of the coupling between a near-surface GaAs/$Al_{0.3}Ga_{0.7}As$ QW, and its free surface, have been explored by Houzay *et al* [55]. Once again, *in situ* PL measurements have been performed on a specific structure, where the sample surface has been exposed to both oxygen and ammonia. As expected, the strength of coupling between the near-surface well and the sample surface has increased on oxidation, and been reduced following ammonia treatment. However, whilst these *in situ* PL measurements seem to argue strongly for the formation of a hybrid state, due to the interaction between the surface and QW states, there has been no additional corroboration, to date, from *ex situ* experiments.

Ex situ PL measurements have been performed on a series of $In_{0.26}Ga_{0.74}As$/GaAs QW structures, where each sample had been grown with a different thickness of GaAs terminating layer [56]. The test structure chosen consisted of 1.1, 3.0, and 5.0 nm wells (growth order), with 100 nm GaAs barriers separating the 3.0 nm QW from the QWs on either side. This allowed the PL intensity ratios, $I_{PL}(50{:}30)$ and $I_{PL}(30{:}11)$, to be used as a measure of the interaction between the near-surface QW and the sample surface, and as a normalising factor between samples, respectively. Figure 6.17 displays the

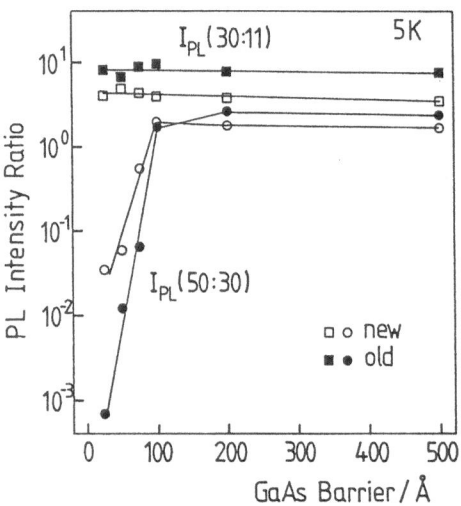

Fig. 6.17. Variation in 5K PL intensity ratios versus GaAs surface-barrier thickness, for a range of $In_{0.26}Ga_{0.74}As$/GaAs QW samples.

dependence, measured at 5K, of $I_{PL}(50:30)$ and $I_{PL}(30:11)$ on GaAs surface-barrier thickness, for a range of samples grown by MBE. The open symbols correspond to measurements on as-grown samples, whilst the filled symbols reflect data obtained from 8 month old samples, where each surface had been dipped briefly in acetone and then blown dry with nitrogen. In both cases, attenuation of $I_{PL}(50:30)$ occurs only for surface-barrier thicknesses of less than 10.0 nm GaAs, with the onset of coupling occurring around 7.5 nm GaAs. It is worth noting that $I_{PL}(30:11)$ remains essentially constant for both as-grown and old samples, thus demonstrating that the intensity ratio $I_{PL}(50:30)$ provides a realistic measure of the degree of interaction between the near-surface QW and the sample surface. At the same time, there was no noticeable red-shift of the PL peak arising from the 5.0 nm near-surface well. A third way in which the thickness of the surface barrier can be varied is to remove material in a series of steps by chemical etching. PL measurements for chemically etched $Al_{0.3}Ga_{0.7}As/GaAs$ QW samples [57], and $In_x Ga_{1-x}As/GaAs$ QW structures [58], have both failed to detect any discernible shift in PL peak energy for the near-surface QW.

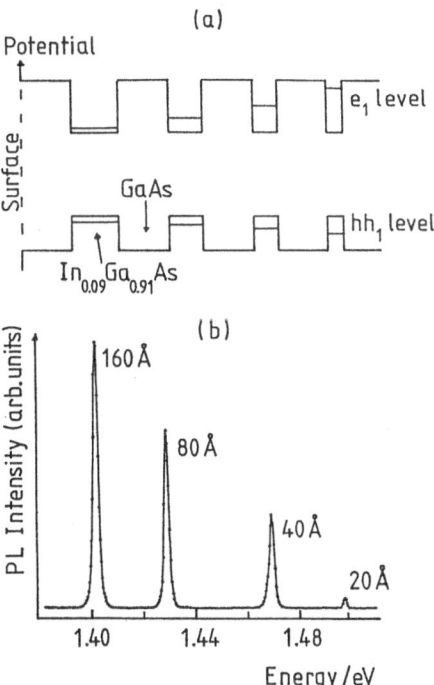

Fig. 6.18. (a) Schematic of the variation in confining potential for $In_{0.09}Ga_{0.91}As/GaAs$ QW samples consisting of (from right) 2, 4, 8, and 16 nm wells, together with (b) typical 10K PL spectrum obtained from such a structure.

A further example of using QWs as a probe of the local electronic environment comes from the work of Sobiesierski *et al* [59]. The primary aim of this study was to improve the quantum efficiency of $In_x Ga_{1-x} As/GaAs$ layers grown on Si substrates, by using post-growth hydrogenation as a means of removing non-radiative recombination centres. Consequently, $In_x Ga_{1-x} As/GaAs$ samples grown directly on GaAs(001) were intended to serve as an additional monitor of the effects of hydrogen implantation. Figure 6.18a provides a schematic of a typical structure, consisting of 2, 4, 8 and 16 nm $In_{0.09} Ga_{0.91} As$ wells separated by 50 nm GaAs barriers. Figure 6.18b displays the corresponding PL spectrum, measured at 10K, from a virgin sample prior to hydrogenation. The compressive strain experienced by the $In_x Ga_{1-x} As$ layers results in only the heavy hole (hh) states being localised with each well. Thus, each PL transition corresponds to radiative recombination between the first electron and heavy hole states (e1-hh1). Since the wells were grown in order of increasing thickness, the 16 nm well, which gives rise to the lowest energy e1-hh1 transition, is situated closest (50 nm) to the sample surface. Hence, the intensity of hydrogen-related features, appearing in PL spectra for hydrogenated samples, should follow the variation of hydrogen content with depth, providing the number of hydrogen-related states remains proportional to the amount of hydrogen present.

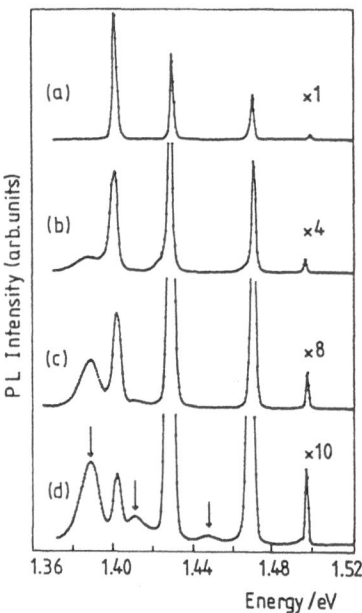

Fig. 6.19. 10K PL spectra for samples consisting of 2.0, 4.0, 8.0 and 16.0 nm $In_{0.09}Ga_{0.91}As/GaAs(001)$ quantum wells, hydrogenated as follows: (a) virgin sample, (b) 6×10^{16}, (c) 3×10^{17} and (d) 2×10^{18} ions cm^{-2}. Arrows indicate energies of hydrogen-related peaks.

Figure 6.19 contains PL spectra, acquired at 10K, from samples exposed to increasing hydrogen dosage. In addition to the four e1-hh1 transitions observed for the virgin sample, a new PL peak appears as a satellite to the 1.402 eV emission from the 16 nm well, at a hydrogen dose of 6×10^{16} hydrogen ions cm^{-2}. As the hydrogen dose increases, two further PL peaks emerge at energies just below the e1-hh1 transition energies of the 8 nm and 4 nm wells. Figure 6.20 shows the PL spectrum obtained from the sample hydrogenated with 2×10^{18} ions cm^{-2}, together with the PL excitation spectrum obtained by monitoring the PL intensity at 1.410 eV, the peak energy of the hydrogen-related satellite in the

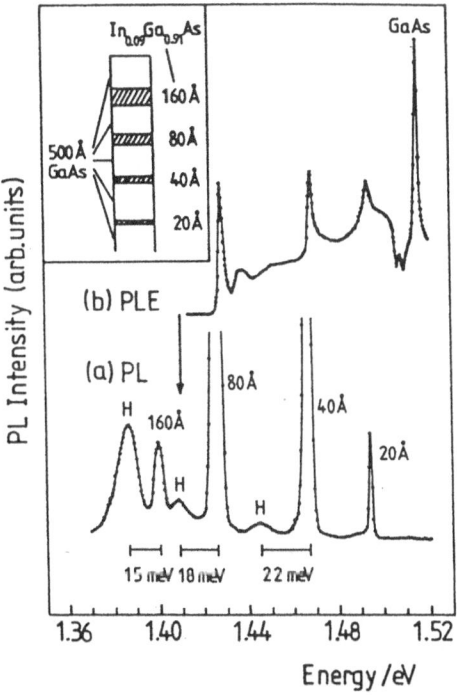

Fig. 6.20. 4.2K PL spectrum obtained from the $In_{0.09}Ga_{0.91}As$/GaAs QW structure shown in the inset, hydrogenated with 2×10^{18} hydrogen ions cm^{-2}, and (b) 4.2K PLE spectrum for the hydrogen-related PL band arising from the 8.0 nm QW.

vicinity of the 8 nm QW emission. The onset of excitation for this hydrogen-related satellite occurs at 1.428 eV, an energy which corresponds exactly to the e1-hh1 transition for the 8 nm well. Likewise, PL excitation spectra taken for the other two hydrogen-related features clearly assign them to radiative transitions occurring within the 16 nm and 4 nm wells. Hence, the sequence in which hydrogen-related PL peaks start to appear indicates that increased hydrogen incorporation leads to the observation of luminescence from hydrogen-related states, located within individual wells, situated at increasing distance from the sample surface. Furthermore, the dependence of the PL spectra on hydrogen dose, coupled with the excitation intensity behaviour of the most heavily hydrogenated samples, seem to indicate [60] that the hydrogen content decreases significantly with increasing distance from the sample surface.

The apparent inhibition of $In_xGa_{1-x}As/GaAs$ QWs to the diffusion of hydrogen has been verified, to some extent, by hydrogenating similar QW structures, which differed [61] only in In alloy composition. PL measurements performed on samples possessing In mole fractions of 0.09, 0.16 and 0.20, revealed that for both 0.16 and 0.20 compositions, shallow hydrogen-related PL arose only from the QW closest to the sample surface. Likewise, hydrogenation [62] of a structure consisting of 6 nm and 8 nm $In_{0.13}Ga_{0.87}As/GaAs$ wells, separated by 350 nm GaAs, has shown that for hydrogen doses up to 10^{18} ions cm^{-2}, implanted at 250°C, hydrogen-related PL is observed only from the 8 nm near-surface QW. However, it is only possible to speak in terms of there being an inhibition to the diffusion of hydrogen, if the hydrogen-related states are deemed to be distributed uniformly throughout each QW region.

Recent measurements [63] have attempted to determine if the hydrogen-related radiative states are indeed indicative of hydrogen bonded within an $In_xGa_{1-x}As$ environment, or whether this luminescence arises solely from hydrogen trapped at the QW interfaces. Secondary-ion-mass-spectroscopy profiles of suitably deuterated $In_{0.2}Ga_{0.8}As$ QW samples have revealed spikes in the deuterium concentration, which correlate precisely with the position of each QW in the structure. The variation in intensity of the deuterium spikes, with increasing distance from the sample surface, tends to suggest that the hydrogen-related radiative recombination centres are associated with an accumulation of hydrogen, in excess of the uniform volume concentration, at the $In_xGa_{1-x}As/GaAs$ QW interfaces. Hydrogenation of GaAs/AlGaAs QWs in a similar manner has not resulted in the observation of any corresponding hydrogen-related PL states. The difference in behaviour of these two materials systems reflects a contrast in interfacial chemistry. The GaAs/AlGaAs heterojunction is predominantly abrupt and, as such, always rough due to the presence of atomic steps in the plane of the interface. On the other hand, the InGaAs/GaAs heterojunction is far less abrupt, but much smoother, due to the inter-mixing which takes place [64, 65], as a result of the surface segregation of In atoms during growth.

6.5.4 Experimental Probing of Quantum Well Eigenstates

Perturbation of the radiative states within a near-surface QW is not the only way in which PL measurements can be used as probes of the local electronic structure. One powerful approach, as demonstrated elegantly by Marzin and Gerard [66], is to incorporate isoelectronic planes of impurities, which then create a short-range potential within the sample. In this particular set of measurements, GaAs/AlGaAs multiple QW samples have been grown, with 1 ML thick planes of either In (attractive potential) or Al (repulsive potential) substituted within each well. The unperturbed sub-band transition energies are determined from PL and PLE measurements, in the absence of any substituted impurity. The introduction of the In or Al planes obviously shifts the PL transition energies. The magnitude of this shift is then measured as a function of the position of the substitutional plane within the well. By performing the appropriate analysis [66], this information can then be used to provide values for the electronic probability density envelope (PDE) of each discernible sub-band. Figure 6.21 shows the remarkable agreement which has been obtained between the experimental PDEs and the $n=1$, $n=2$ and $n=3$ electron sub-bands calculated within the effective-mass approximation.

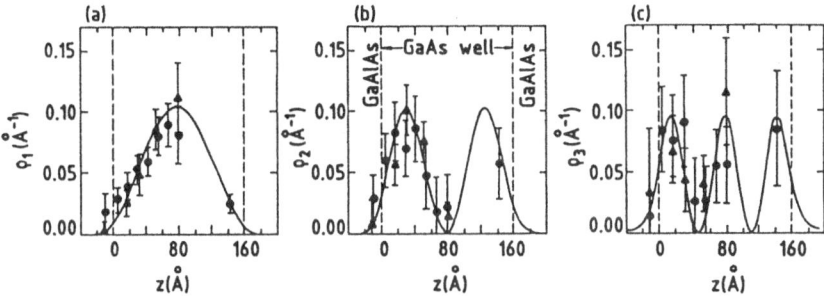

Fig. 6.21. Experimental probability density envelopes, ρ, for (a) $n=1$, (b) $n=2$, and (c) $n=3$ electron energy levels, compared to the theoretical effective-mass result. Dots and triangles are experimental results [66].

However, it needs to be pointed out that such agreement between the predicted and experimental optical properties is never sufficient proof, in itself, that the heterojunction interfaces are abrupt in reality. A case in point is the $In_xGa_{1-x}As$/GaAs system where, even in the presence of good agreement between the theoretical predictions and the corresponding experimental measurements, recent data obtained by both secondary-ion mass spectroscopy (SIMS) [64], and *in situ* photoelectron spectroscopy [65], have shown considerable intermixing exists at the $In_xGa_{1-x}As$/GaAs interface, due to the surface segregation of In atoms during growth.

References

1. Lenard, P., Schmidt, F., Tomaschek, R.: Handbuch der Experimental Physik vol.23, parts 1 and 2. Akademische Verlagsgesellschaft, Leipzig, 1928
2. Hilsch, R.: Zeitschrift für Physik *44*, 860 (1927)
3. Smakula, A.: Zeitschrift für Physik *46*, 558 (1928)
4. Hilsch, R., Pohl, R.W.: Zeitschrift für Physik *48*, 384 (1928)
5. Schleede A., Gantzckov, H.: Zeitschrift für Physikalische Chemie 106, 37 (1923)
6. Leverenz, H.W. in:: Fonda, G.R., Seitz, F. (eds.) Solid Luminescent Materials. Wiley, New York 1948, p. 148
7. See, for example, Lax, M.: Journal of Chemical Physics *20*, 1752 (1952)
8. Dexter, D.L., Knox, R.S.: Excitons. Interscience Tracts on Physics and Astronomy. Wiley, New York 1965.
9. Cooley, J.W., Tukey, J.W.: Mathematical Computing *19*, 297 (1965)
10. Curie, D.: Luminescence in Crystals. Methuen, London 1963
11. Klick, C.C., Schulman, H. in: Seitz, F., Turnbull, D., (eds.) Solid State Physics vol. 5. Academic Press, New York 1957, p. 100
12. Thomas, D.G., Gershenzon, M., Trumbore, F.A.: Physical Review *133 A*, 269 (1964)
13. Reynolds, D.C., Litton, C.W., Park Y.S., Collins, T.C.: Journal of the Physical Society of Japan *21*, supp. 143 (1966)
14. Ryan, F.M., Miller, R.C.: Physical Review *148*, 858 (1966)
15. Herman, M.A., Bimberg D., Christen, J.: Journal of Applied Physics *70*, R1 (1991)
16. Kopf, R.F., Schubert, E.F., Harris T.D., Becker, R.S.: Applied Physics Letters 58, 631 (1991)
17. Singh J., Bajaj, K.K.: Journal of Applied Physics *57*, 5433 (1985)
18. Lee S.M., Bajaj, K.K.: Applied Physics Letters *60*, 853 (1992)
19. Aspnes, D.E.: Surface Science *132*, 406 (1983)
20. Ellis, B., Moss, T.S.: Solid-State Electronics *13*, 1 (1970)
21. Nelson, R.J., Sobers, R.G.: Applied Physics Letters *32*, 761 (1978)
22. Nelson, R.J.: Journal of Vacuum Science and Technology *15*, 1475 (1978)
23. Nelson, R.J., Williams, S.J, Leamy, H.J., Miller, B., Casey,H.C., Parkinson, B.A., Heller, A.: Applied Physics Letters *36*, 76 (1980)
24. Offsey, S.D., Woodall, J.M., Warren, A.C., Kirchner, P.D., Chappell, T.I., Pettit, G.D.: Applied Physics Letters *48*, 475 (1986)
25. Yablonovitch, E.Skromme, B.J., Bhat, R., Harbison J.P., Gmitter,T.J.: Applied Physics Letters *54*, 555 (1989)
26. Olego, D.J.: Applied Physics Letters *51*, 1422 (1987)
27. Chang, R.R., Iyer, R., Lile, D.L.: Journal of Applied Physics *61*, 1995 (1987)
28. Iyer, R., Lile,D.L.: Applied Physics Letters *59*, 437 (1991)
29. Hall, R.N.: Physical Review *87*, 387 (1952)
30. Shockley, W., Read, W.T.: Physical Review *87*, 835 (1952)
31. Saitoh, T., Iwadate, H., Hasegawa, H.: Japanese Journal of AppliedPhysics *30*, 3750 (1991)
32. Scharfetter, D.L., Gummel, H.K.: IEEE Transactions on Electronic Devices *16*, 64 (1969)
33. Street, R.A., Williams R.H., Bauer, R.S.: Journal of Vacuum Science and Technology *17*, 1001 (1980).
34. Obersta, J.D., Streetman, B.G.: Surface Science *108*, L470 (1981)
35. Kim, T.S., Lester, S.D., Streetman, B.G.: Journal of Applied Physics *61*, 2072 (1987)
36. Sobiesierski, Z., Dharmadasa, I.M., Williams, R.H.: Applied Physics Letters *53*, 2623 (1988)

37. Dharmadasa, I.M., Thornton J.M., Williams, R.H.: Applied Physics Letters *54*, 137 (1989)
38. Mclean, A.B., Dharmadasa, I.M., Williams, R.H.: Semiconductor Science and Technology *1*, 137 (1986)
39. Sandroff, C.J., Turco-Sandroff, F.S., FlorezL.T., Harbison, J.P.: Journal of Applied Physics *70*, 3632 (1991)
40. Davies, G.J., Heckingbottom, R., Ohno, H., Wood, C.E.C., Calawa, A.R.: Applied Physics Letters *37*, 290 (1980)
41. Moison, J.M., Bensoussan, M., Houzay, F.: Physical Review *B 34*, 2018 (1986)
42. Hollinger, G., Gallet, D., Gendry, M., Santinelli, C., Viktorovitch,P.: Journal of Vacuum Science and Technology *B 8*, 832 (1990)
43. Viktorovitch, P., Gallet, D., Gendry, M., Hollinger, G., Schohe, K., Benyatttou, T., Tabata, A., Regaud, D., Guillot, G.: Second International Conference. InP and Related Materials, Denver 1990. IEEE, New York 1990, p. 148
44. Cohen, R.M., Kitamura, M., Fang, Z.M.: Applied Physics Letters *50*, 1675 (1987)
45. Yablonovitch, E., Cox, H.M., Gmitter, T.J.: Applied Physics Letters *52*, 1002 (1988)
46. Kanski, J., Nilsson, P.O., Karlsson, U.O., Svensson, S.P. in: Anastassakis, E.M., Joannopoulos, J.D. (eds.) Proceedings of the 20th International Conference on the Physics of Semiconductors, Thessaloniki 1990. World Scientific, Singapore 1990, p. 195
47. Sobiesierski, Z., Clark, S.A., Williams, R.H., Tabata, A., Benyattou, T., Guillot, G., Gendry, M., Hollinge, G., Viktorovitch, P.: Applied Physics Letters *58*, 1863 (1991)
48. Seifert, W., Fornell, J.O., Ledebo, L.A., Pistol, M.E., Samuelson, L.A: Applied Physics Letters *56*, 1128 (1990)
49. Hollinger, G., Gendry, M., Duvault, J.L., Santinelli, C., Ferret, P., Miossi, C., Pitaval, M.: AppliedSurface Science *56-58*, 665 (1992)
50. Schneider, R.P., Wessels, B.W.: Applied Physics Letters *57*, 1998 (1990)
51. Sobiesierski, Z., Clark, S.A., Williams, R.H.: AppliedSurface Science *56-58*, 703 (1992)
52. Moison, J.M., Elcess, K., Houzay, F., Marzin, J.Y., Gerard, J.M., Barthe, F., Bensoussan, M.: Physical Review *B 41*, 12945 (1990)
53. Van der Veen, J.F., Larsen, P.K., Neave, J.H., Joyce, B.A.: Solid State Communications *49*, 659, (1984)
54. Sauvage-Simkin, M., Pinchaux, R., Massies, J., Calverie, P., Jedrecy, N., Bonnet, J., Robinson, I.K.: Physical Review Letters *62*, 563 (1989)
55. Houzay, F., Moison, J.M., Elcess, K., Barthe, F.: Superlattices and Microstructures *9*, 507 (1991)
56. Sobiesierski, Z., Westwood, D.I.: Superlattices and Microstructures 12, 267 (1992)
57. Chang, Y.L., Tan, I.H., Mirin, R., Zhang, Y.H., Merz J.L., Hu, E.: Institute of Physics Conference Series *129*, 281 (1992)
58. Sobiesierski, Z., Westwood, D.I., Woolf, D.A., Fukui, T., Hasegawa, H.: Journal of Vacuum Science and Technology *B 11*, 1723 (1993)
59. Sobiesierski, Z., Woolf, D.A., Westwood, D.I., Frova A., Coluzza, C.: Solid State Communications *81*, 125 (1992)
60. Sobiesierski, Z., Woolf, D.A., Frova, A., Phillips, R.T.: Journal of Vacuum Science and Technology *B 10*, 1975 (1992)
61. Sobiesierski, Z., Woolf, D.A., Westwood, D.I.: Superlattices and Microstructures *12*, 261 (1992)
62. Chang, Y.L., Tan, I.H., Hu, E., Merz, J., Frova A., Emiliani, V.: Journal of Vacuum Science and Technology *B 11*, 1702 (1993)
63. Sobiesierski, Z., Clegg, J.B.: Applied Physics Letters *63*, 926 (1993)
64. Muraki, K., Fukatsu, S., Shiraki, Y., Ito, R.: Journal of Crystal Growth *127*, 546 (1993)

65. Larive, M., Nagle, J., Landesman, J.P., Marcadet, X., Mottet, C., Bois, P.: Journal of Vacuum Science and Technology *B 11*, 1413 (1993)
66. Marzin,J.Y., Gerard, J.M.: Physical Review Letters *62*, 2172 (1989)
67. Canham, L.T.: Applied Physics Letters *57*, 1046 (1990)
68. Lehmann, V., Gosele, U.: Applied Physics Letters *58*, 856 (1991)
69. Gardelis, S., Rimmer, J.S., Dawson, P., Hamilton, B., Kubiak, R.A., Whall, T.E., Parker, E.H.C.: Applied Physics Letters *59*, 2118 (1991)
70. Isshiki, H., Saito, R., Kimura, T., Ikoma, T.: Journal of AppliedPhysics *70*, 6993 (1991)

Chapter 7. On the Theory
of Second Harmonic Generation

Andrea d'Andrea,[1] Michele Cini,[2] Rodolfo Del Sole,[2] Lucia Reining,[3] Claudio Verdozzi,[4] Raffaello Girlanda,[5] Edoardo Piparo,[5] David Hobbs,[6] and Denis Weaire[6]

[1] I.M.A.I., C.N.R., Casella Postale 10, I-00016 Roma, Italia

[2] Dipartimento di Fisica, II Università di Roma, 'Tor Vergata', Via della Ricerca Scientifica, I-00133 Roma, Italia

[3] Centre Européen de Calcul Atomique et Moléculaire, Bât. 506, Université Paris Sud, Orsay, France

[4] Interdisciplinary Research Centre in Surface Science, The University of Liverpool, P.O. Box 147, Liverpool L69 3BX, UK

[5] Istituto di Struttura della Materia, Facoltà di Scienze m.f.n.,Università di Messina, Sant'Agata - Messina, Italia

[6] Department of Physics, Trinity College, Dublin 2, Ireland

In Sect. 1.5, second-harmonic and sum-frequency generation (SHG and SFG) at surfaces and interfaces was introduced. Experimental results are discussed in Chapter 8. Various aspects of the theory of SHG at surfaces and interfaces are discussed in this chapter.

7.1 General Theory of Second Harmonic Generation

Current understanding [1, 2] of SHG is still essentially based on the lowest-order perturbation expansion in the electron-photon coupling, and on the independent-electron scheme. This theoretical approach, which is oversimplified in principle, already involves heavy computational effort. It still deserves considerable attention, however, because we need to develop its implications for complex systems, like surfaces and interfaces; very little such work has been reported to date.

Our recent contributions [3, 4] essentially belonged to this line of thought. We aimed at the inclusion of self-consistency, which is needed in interface problems, as least as far as it is represented by Fresnel's equations. To this end, the SH intensity was obtained in terms of the photon Green's function. The approach is reviewed in Appendix A.1; it was used for the slab calculation of Sect. 7.2, where the A^2 term was neglected for the sake of simplicity.

Next, we wish to propose a new theory of the nonlinear response, which is based on the method of Excitation Amplitudes [5] and is therefore non-perturbative. The point is that very weak effects, that could not be included by the usual methods and were previously neglected, are now coming to the foreground. Both experiment and the above mentioned calculations show that the ratio of incident to SH photon fluxes can be an extremely large number, such as we could expect to meet in astrophysics rather than surface physics. The new method can deal with small effects because it produces exact solutions.

We set up a formalism (Appendix A.2) describing the scattering of a laser beam by a material system in the dipole approximation. We neglect the tiny broadening effects due to spontaneous decay (interaction with vacuum photons), but include the laser beam-matter interaction to all orders.

The method of Excitation Amplitudes leads to exact, closed form solutions, both in the incident and scattered field. However, the SH field is very weak, and we limit ourselves to first order in the emitted photon. In Appendix A.3 we review the physical meaning of this approach and the expansion of the transition amplitude in Excitation Amplitudes. At this point, we are able, in principle, to compute the nonlinear repsonse of the system as a function of the duration, t, of the beam-matter interaction. Since we shall be primarily interested in the long term behaviour, a direct application of the above results should be supplemented by an asymptotic limit ($t \to \infty$). This would require the knowledge of the analytic structure of the Fourier transform results, which is not easily obtained by direct computation. Fortunately, we were able to derive formulae that give the asymptotic behaviour directly. Deferring these developments to Appendix A.4, here we concentrate on the physical results for a two-level system.

7.1.1 The Many-Photon Approach: A Simple Example

Although the theory applies to any system, and should allow realistic calculations, here we consider a two-level model, with

$$H_e = \varepsilon_1 n_1 + \varepsilon_2 n_2 \tag{7.1}$$

$$H_I = (a_1^\dagger a_2 + a_2^\dagger a_1) g (b + b^\dagger) \tag{7.2}$$

$$H'_I = (a_1^\dagger a_2 + a_2^\dagger a_1)g(b' + b'^\dagger) \tag{7.3}$$

Here, ε_m are the one-electron energy levels and n_m are the occupation number operators. The system is initially in the state $a_1^\dagger |0>$, where $|0>$ is the vacuum for both the b and b' photons. The energy levels are taken to be $\varepsilon_1 = 0$ and $\varepsilon_2 = 1$, that is, all energies are measured in units of the fundamental excitation energy. The matter-radiation coupling is adiabatically switched off ($g \propto e^{-\Gamma t}$, with $\Gamma \to 0$), which is a well known device to ensure that the theory is well behaved for $t \to \infty$.

First, we use the above formalism to investigate the dynamical Stark effect, which takes place when a coherent field is resonantly tuned to a two-level system. Schuda et al [6], using a dye laser and a beam of Na atoms, demonstrated experimentally that the scattered radiation shows a splitting into three components. The laser was tuned to the transition from the F=2 hyperfine component of the $3^2 S_{1/2}$ ground level to the F=3 component of $3^2 P_{3/2}$. The laser was so monochromatic that there was no contamination from the F=2 component of $3^2 P_{3/2}$, and the selection rules ensured no mixing with other states, so that a genuine two-level system was observed. The three-peaked structure of the spectrum can be understood by a simple approach [7] based on lowest-order perturbation theory and the rotating wave approximation. The calculated spectrum consists of three Lorentzians, with intensity 1:2:1, separated by $2|g\beta|$, where β is the eigenvalue of b in the coherent state. There is also an overall shift which results from level renormalisation, and is independent of the exciting field. The rotating wave approximation was also used by Carmichael and Walls [8], who considered the master equation for the reduced density matrix in the Markoffian approximation.

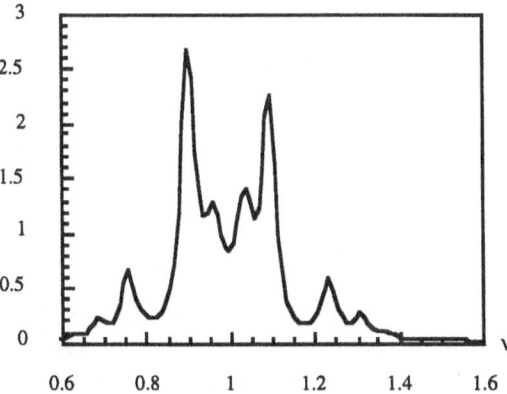

Fig. 7.1 Scattered radiation intensity as a function of frequency, assuming $\beta = 1$ and $g = 0.1$ [9].

Our exact quantum mechanical solution supports the earlier approximate treatments for $\beta \to \infty$, but also reveals corrections which become important at small enough β. For example, in Fig. 7.1 we show $P(\infty)$ in the resonant case ($\omega_0 = 1$), assuming $\beta = 1$ and $g = 0.1$. Good convergence was achieved by including the poles of excitation amplitudes up to $\omega_{max} = 6.3$, and excitation amplitudes involving up to 18 photons. The line shape does not conform to the expectations of the elementary theory, and looks fairly complicated. Note that according to the elementary theory, the spectrum should depend on g and β through their product. This is not borne out by the exact results.

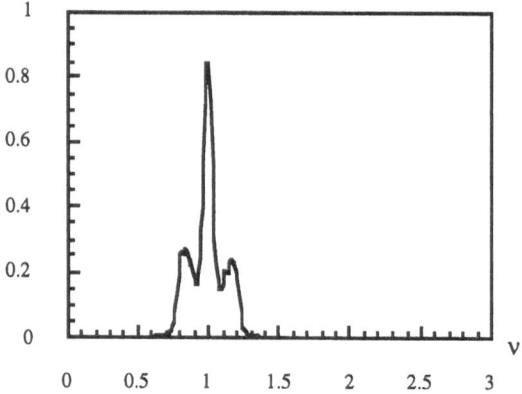

Fig. 7.2. Scattered radiation intensity as a function of frequency, assuming $\beta = 2$ and $g = 0.05$ [9].

In Fig. 7.2, we show $P(\infty)$, still in the resonant case ($\omega_0 = 1$), but assuming $\beta = 2$ and $g = 0.05$. Good convergence was achieved by including the poles ofexcitation amplitudes up to $\omega_{max} = 5.8$ and excitation amplitudes involving up to 24 photons. The shape of the spectrum is completely different from the previous case. There are still many features, but their envelope now recalls the three-peaked structure which is familiar from the previous treatments. True quantum fluctuations arise when the dynamical Stark effect depends on g and β separately, and not just on the product, $g\beta$.

To see the relevance of the above theory to SHG, let us now consider the same two-level system tuned by laser light having a frequency $\omega_1 = 0.5$, $g = 0.1$, $\Gamma = 0.02$ (in units of the level separation), and $\beta = 1$. We now obtain an elastic peak, due to Rayleigh scattering, at $v = 0.5$, followed by an intense SH satellite, close to the natural frequency of the two-level system (Fig. 7.3). One can notice, however, that the satellite shows a noticeable shift to higher frequency and shows an interesting Mollow-like three-peak structure, similar to the dynamical Stark effect. This is a genuine many-photon phenomenon that could not be obtained from the previous treatments based on perturbation theory. We should like to stress that it would disappear altogether if we adopted the rotating-wave approximation, as in conventional treatments.

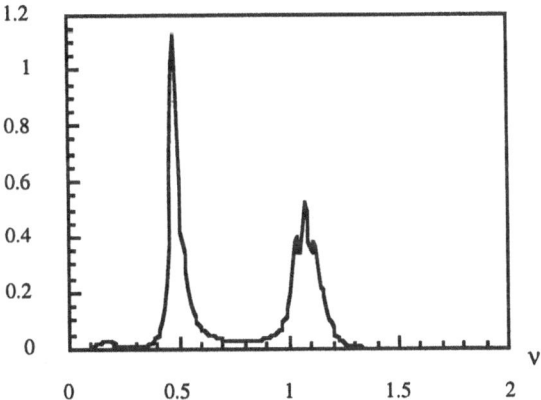

Fig. 7.3. Scattered radiation intensity as a function of frequency, assuming $\omega_1 = 0.5$, $g = 0.1$, $\Gamma = 0.02$, and $\beta = 1$ [9].

The spectrum still depends on g and β separately, but we can show that the SH feature persists for any β. We found that the limit $g \to 0$, $\beta \to \infty$ $g\beta \to$ constant, corresponds to a semiclassical treatment of the field and can be calculated exactly in our the quantum mechanical treatment without the rotating-wave approximation. Details of both the quantum and semiclassical formalisms are published elsewhere [9]. New sum rules for nonlinear optics [10] have also been obtained from the above theory.

7.2 Slab Calculations

In this Section we will show how it is possible to carry out realistic calculations of SHG at surfaces based on the lowest-order perturbation expansion in the electron-photon coupling.

In principle, we should compute the non-local, first- and second-order susceptibilities for a semi-infinite crystal, and then solve the coupled equations for light propagation. The difficulty of this task is illustrated by the fact that the calculation of the *linear* non-local susceptibility, $\chi^{(1)}(z, z'; \omega)$, as a function of z and z' has not yet been done for a semiconductor slab because of the computational complexity (see Sect. 2.2). The method described in Appendix A.1 bypasses these difficulties and makes surface SHG calculations feasible. The emitted light intensity, at frequency 2ω, is directly calculated according to third-order perturbation theory, by-passing the calculation of the non-local $\chi^{(2)}$ and the solution of Maxwell's equations. In this Section we show how to calculate the SHG yield of a semiconductor surface according to this formalism, using the tight-binding method. The light propagation near the surface is described according to the anisotropic three-phase model [11] (see Sect. 2.2), up to zero-order in d/λ, where d is the depth of the surface layer and λ is the wavelength of the incident EM radiation. The full equations may be found in Appendix A.

We take Si(111)1x1-As as a test case basically for three reasons: (i) the atomic and electronic structure of this surface are well known [12]; (ii) the absence of reconstruction leads to a small surface unit cell, whose size affects, in a drastic way, the computer time needed; (iii) SHG experiments have been carried out on this surface by Kelly *et al* [13].

The atomic and electronic structure of the Si(111)1x1-As surface is presently well understood [12, 14]. Arsenic atoms substitute Si atoms in the first layer, giving rise to doubly occupied As dangling bonds. Thus the inherent instability of the ideal Si(111) surface is removed, because the unpaired electrons of that surface are now coupled into a stable lone pair. We have computed the electronic structure of a slab of 12 Si(111) layers, where the first and last layer Si atoms are substituted by As atoms, according to the ight-binding method described in Ref. 15 and in Sect. 2.2. The outward relaxation of the surface layers has been properly included.

Some problems arise in the calculation of the response functions when a semi-infinite crystal is mimicked by a slab. The point is that a slab symmetrical with respect to its central plane has a spurious inversion symmetry which is not present in the semi-infinite sample. As a consequence, some response functions might be zero. This occurs, for instance, in the case of the xz elements of the dielectric tensor, which would be zero in a symmetrical slab, even if they are non-vanishing for a single surface [16]. This problem has been solved by calculating the integral of the xz dielectric susceptibility, which yields the xz component of the slab dielectric tensor, over only half of the slab [16]. A similar problem arises in the case of SHG calculations: a symmetrical slab of Si will have inversion symmetry, so that will not yield a SHG. Why does a slab yield no SHG, while a single surface does? Because there is destructive interference, of course, between the waves emitted from the two surfaces. Therefore, even though the slab is not symmetric and yields some SHG, it is not representative of the single surface response.

In order to cope with this problem, we use the following trick. We assume that the electric field of the incident light has a z-dependence, $S(z)$, such that it is constant ($S(z) = 1$) on front-surface active layer, and vanishes ($S(z) = 0$) on the corresponding layer of the back surface. The big change of $S(z)$ occurs in the intermediate "bulk" region, where no SHG occurs. Another requirement on $S(z)$, due to our tight-binding method, is that it must be smooth on the atomic scale, so that we can neglect its intersite matrix elements.

We calculate the probability of SHG in the gauge where the scalar potential is not present. In this gauge the interaction between the electrons and the incident wave can be written as:

$$V_A = i(e/2m\omega)[S(z)\mathbf{E}^s \cdot \mathbf{p} + \mathbf{p} \cdot \mathbf{E}^s S(z)] \tag{7.4}$$

neglecting the term proportional to A^2. Here $\mathbf{E}^s$ is the electric field of the incident light at the surface. Therefore one has to use the operator

$$p = [S(z)\mathbf{p} + \mathbf{p}S(z)]/2 \tag{7.5}$$

rather than $\mathbf{p}$ in the two right-hand matrix elements in equations B.7, B.7a and B.13, which refer to the incident light. Since the wave functions are expanded in local orbitals, we need to calculate the matrix elements of $\mathbf{p}$ and p between orbitals n and n' in cells $\mathbf{R}$ and $\mathbf{R}'$, respectively. As in the case of the work on linear optical properties [11], we calculate these matrix elements using the identity

$$\mathbf{p} = i(m/\hbar)[\mathbf{H}, \mathbf{r}] \tag{7.6}$$

and neglecting the matrix elements of the position operator, $\mathbf{r}$ (and of $S(z)$), between different atoms. The intra-atomic matrix elements are obtained (Sect. 2.3.1) by a fit of bulk Si optical properties, as $<s|x|p_x> = 0.027$ nm and $<s^*|x|p_x> = 0.108$ nm. It is straightforward to obtain

$$<n\mathbf{R}|p|n'\mathbf{R}'> = i(m/\hbar)\{[S(z_{n\mathbf{R}}) + S(z_{n'\mathbf{R}'})]/2\}\sum_{n''}<n\mathbf{R}|\mathbf{H}|n''\mathbf{R}'><n''\mathbf{R}'|\mathbf{r}|n'\mathbf{R}'>$$
$$- <n\mathbf{R}|\mathbf{r}|n''\mathbf{R}><n''\mathbf{R}|\mathbf{H}|n'\mathbf{R}'> \tag{7.7}$$

where $z_{n\mathbf{R}} = <n\mathbf{R}|z|n\mathbf{R}>$ is the z-component of the centre-of-charge of the $n\mathbf{R}$ orbital. The matrix elements of $\mathbf{p}$ are obtained from equation 7.7 by putting $S(z) = 1$.

In our calculations we use

$$S(z) = \cos^2(\pi z/2L) \tag{7.8}$$

where $z = 0$ and $z = L$ are the positions of the first and last atomic (As) layer respectively. We find that the results very weakly depend on the particular choice of $S(z)$.

Because of the 3m symmetry of the Si(111)1x1-As surface, we need to calculate only four independent components of X_{ijk}, namely X_{xxx}, X_{zzz}, X_{zxx} and X_{xxz}, where x and y are along the principal axes of the surface, namely the [11$\bar{2}$] and [1$\bar{1}$0] directions, respectively. In general, the plane of incidence $x'z$

will be at an angle, ψ, with respect to the x-axis. In this case, *anisotropic* components change according to

$$X_{x'x'x'} = (\cos^3 \psi - 3\sin^2 \psi \cos \psi)X_{xxx} \qquad (7.9a)$$

$$X_{y'y'y'} = (3\sin \psi \cos^2 \psi - \sin^3 \psi)X_{xxx} \qquad (7.9b)$$

$$X_{x'y'y'} = (3\sin^2 \psi \cos \psi - \cos^3 \psi)X_{xxx} \qquad (7.9c)$$

$$X_{y'x'x'} = (\sin^3 \psi - 3\sin \psi \cos^2 \psi)X_{xxx} \qquad (7.9d)$$

All other elements do not depend on ψ. They equal the corresponding components with x and y in place of x' and y', respectively.

We evaluate the real part of $X_{ijk}(\omega)$ according to equation B.13, and then the imaginary part according to equation B.14. Each sum in equation B.13 is over the bands and the wave vectors, $\mathbf{k}$ (parallel to the surface). Because of the momentum conservation rule, a single sum over $\mathbf{k}$ is involved. It is carried out using 126 special points in the irreducible part of the hexagonal SBZ. We calculate $X_{ijk}(\omega)$ for $\hbar\omega < 2$ Rydberg, which is therefore the higher limit of the integral in equation B.14. A single run takes about 30 minutes of *Cray-YMP* CPU time. Calculations with a cut-off of 1 Rydberg show good convergence in the SHG intensity for $\hbar\omega < 5$ eV.

The intensity of the SHG for incident and outgoing p-polarised light is shown by the full line in Fig. 7.4. The angle of incidence is $60°$, and the azimuthal angle, ψ, is zero: the plane of incidence includes the $[11\bar{2}]$ direction. It is evident that the strongest SHG occurs for $\hbar\omega$ in the range of linear-absorption band, around 4.5 eV. The SHG is much lower in the experimentally accessible range, below 2.5 eV. An important point is that of the relevance of resonances at ω or at 2ω. Equation B.13 can be decomposed in a 1ω contribution and a 2ω one, both shown in Fig. 7.4. The 2ω resonances dominate the lower energy part of the curve, which is of experimental interest, while both of them affect the higher-energy part, around 5 eV, with a somewhat greater relevance for 1ω transitions. An interesting point in Fig. 7.4 is the apparent divergence of the generated intensity as ω goes to zero. Actually, the SHG must vanish as ω goes to zero, as a consequence of the generalised f-sum rule [17]; the weak divergence of the solid curve (the sum of the two contributions) is due to the finite extension of the Kramers-Krönig transform (equation B.14). However, the divergence becomes dramatic for the separate contributions. In fact, both of them must be included to satisfy the generalised f-sum rule.

An important issue to be addressed is the capability of SHG to be a spectroscopic tool for surfaces. The intensity (see equations B.9 and B.10)

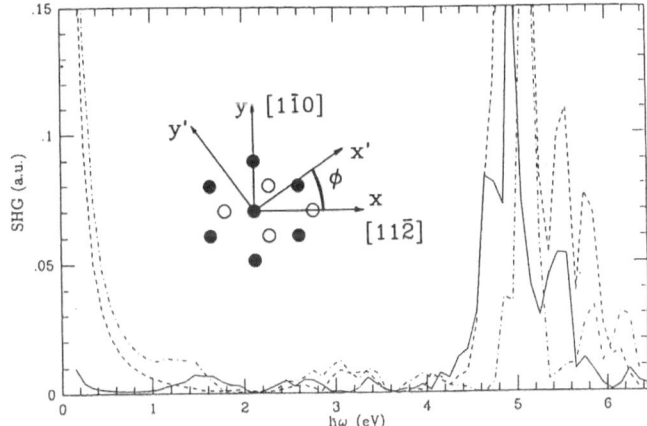

Fig. 7.4. Calculated SHG from Si(111)1x1-As. The ratio of the emitted intensity to the square of the incident intensity, $I(2\omega)/I_0^2(\omega)$, is plotted. The plane of incidence contains the [11$\bar{2}$] direction and the angle of incidence is 60°. The incident and outgoing light are p-polarised. Dashed line: contribution of 1ω resonances. Dot-dashed line: contribution of 2ω resonances. Full line: full contribution [18].

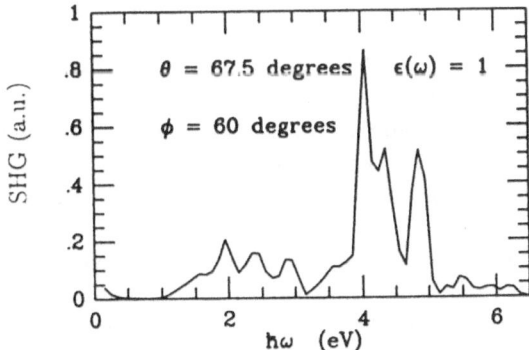

Fig. 7.5. Calculated SHG from Si(111)1x1-As. The ratio of the emitted intensity to the square of the incident intensity, $I(2\omega)/I_0^2(\omega)$, is plotted. The plane of incidence is at an azimuthal angle, $\psi = 60°$ relative to the [11$\bar{2}$] direction, and the angle of incidence, ϕ_0, is 67.5°. The incident and outgoing light are p-polarised. Light propagates as in vacuum ($\varepsilon(\omega) = 1$ everywhere) [18].

depends not only on $X_{ijk}(\omega)$, which involves the surface spectrum, but also on the Fresnel factors of the incoming and out coming waves, which depend strongly on the dielectric function. Therefore the question arises of how strongly the energy dependence of the latter affects the SHG. In order to understand this point, we have made a calculation using $\varepsilon(\omega) = 1$ in the Fresnel factors. In this way, the light propagation is as in vacuum, and the lineshape is uniquely determined by $X_{ijk}(\omega)$. It can be easily shown that, in this case, the SH intensity is proportional to $|X(\omega)|^2/\omega^4$. The result of the calculation is shown in Fig. 7.5, and should be compared with the full line in Fig. 7.4, which takes $\varepsilon(\omega)$ into account. A dramatic increase of intensity is apparent, arising from the following effect: the dielectric function screens the surface field, so that when it is unity the latter is largest. Also, the low-energy part of the curve, below 3 eV, is qualitatively similar to that in Fig. 7.4, so that we can say that SHG spectroscopy in the range of experimental interest yields information on the surface spectrum. However, a quantitative interpretation must rely on calculations accounting for light propagation.

The dielectric function has important effects on the higher-energy part of the spectrum, above 3 eV. In particular, the structures between 5 and 6 eV are hardly affected by $\varepsilon(\omega)$, since it is not large in this range. In contrast, the structures between 4 and 5 eV are severely reduced by the presence of $\varepsilon(\omega)$, which has its maximum in this region.

A last point to be checked is the dependence on the spatial variation of the dielectric function near the surface. This is not included in our model, which assumes an abrupt termination of $\varepsilon(\omega)$ at the surface. However, it can be easily incorporated in the spirit of the three-phase model, discussed in Sect. 1.3 and 2.2 [11, 19]. We assume that there is a surface layer of depth, d, with dielectric function $\varepsilon(\omega)$, as defined in Sect. 1.3 (in this chapter $\varepsilon(\omega)$ has, so far, been the bulk dielectric function, which we now call $\varepsilon_b(\omega)$). To the zeroth order in d/λ, the electric field component parallel to the surface and the electric-displacement component, D_z, are constant across the surface layer. Therefore we can account for it by simply changing E_z^s into $D_z/\varepsilon(\omega) = E_z^s \varepsilon_b(\omega)/\varepsilon(\omega)$. The result of this calculation is shown in Fig. 7.6. The SHG intensity is surprisingly stable below 4.5 eV, but there are large changes above, where the dependence on the dielectric function is important.

We study now the angular dependence of the SHG intensity. We choose an angle of incidence of 67.5°, as in the experiment [13], and make plots for several values of ψ. The results are shown in Fig. 7.7 for p-polarised incident and outgoing light ($p \rightarrow p$). It is apparent from the plots for the other polarisations (not shown here) that a decrease of the peak intensity from $p \rightarrow p$ to $p \rightarrow s$ to $s \rightarrow p$ to $s \rightarrow s$ polarisations occurs. This is related to the fact that the components of X_{ijk} involving z-values of i, j or k are larger than the other ones, since the breaking of the inversion symmetry is associated with the z-axis.

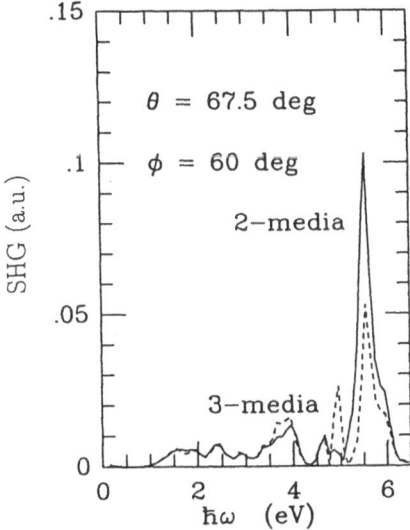

Fig. 7.6. Calculated SHG from Si(111) 1x1-As. The ratio of the emitted intensity to the square of the incident intensity, $I(2\omega)/I_0^2(\omega)$, is plotted. The azimuthal angle, ψ, is 60°, and the angle of incidence, ϕ_0, is 67.5°. The incident and outgoing light are p-polarised. Full line: two-phase model. Dashed line: three-phase model [18].

The $p \rightarrow p$ spectrum plotted in Fig. 7.7 demonstrates the interplay of X_{xxx}, which yields angular-dependent contributions, and the ψ-independent X_{zzz}, X_{zxx} and X_{xxz} components. Inspection of equations 7.9 shows that the structure around 5.5 eV, which is only weakly ψ-dependent, originates from the latter terms. Since no surface structure is present in the linear spectrum at 5.5 or 11 eV, we believe that it originates from the minimum of $|\varepsilon(\omega)|$ which is near $\omega = 5.2$ eV in our calculations. On the other hand, the structures just below 5 eV, near 4 eV, and between 1.5 eV and 2 eV, are due to the interplay of x and z polarisations, in agreement with the linear spectrum of Fig. 2.4.

   The dependence on the angle of incidence of the $p \rightarrow p$ spectrum, with $\psi = 60°$, is shown in Fig. 7.8. It is clear from this figure that the angle of incidence does not modify the spectrum qualitatively. The long-dashed line, calculated for $\phi_0 = 67.5°$, should be directly compared to the experiment [13]. The lineshape is qualitatively similar, but the experimental intensity is about 30 times smaller. The reason of such discrepancy is unclear. It should not be due to our assumption of an abrupt truncation of the bulk dielectric constant, since the calculation carried out according to the three-phase model, shown in Fig. 7.6, yields negligible differences in the energy range of interest. From the theoretical side, a possible explanation might be our neglect of the A^2 term in the calculations of the SHG amplitude. It would be nice to include this term *via* the self-consistent procedure outlined in equation A.2 of Appendix A.

In conclusion, the SHG intensity does not depend very much on the contents of the calculation in the low-energy experimentally accessible range. It mostly reflects the structure of the surface spectrum contained in $X(\omega)$ and, therefore, appears to be a promising tool for surface spectroscopy.

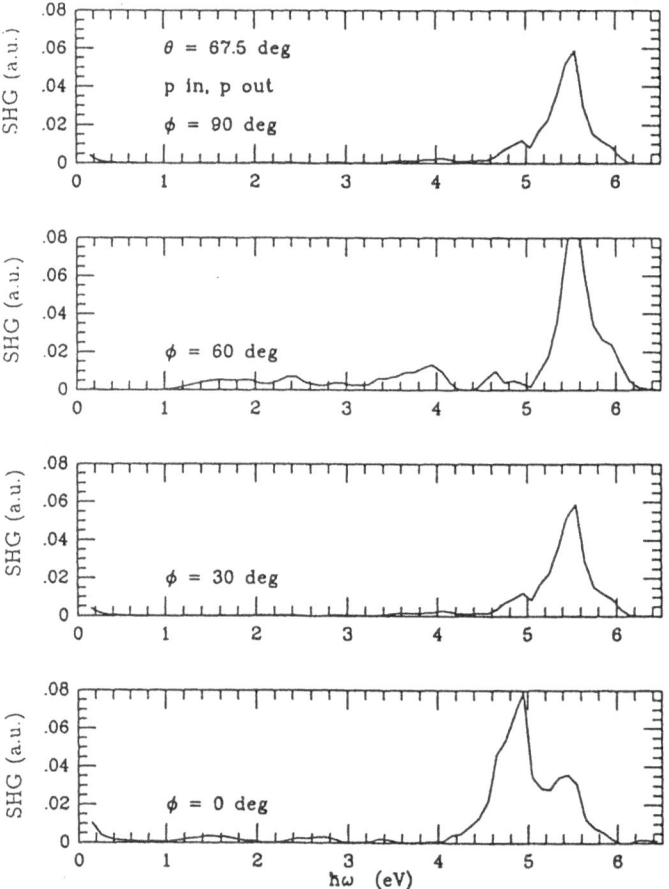

Fig. 7.7. Calculated SHG from Si(111)1x1-As for various azimuthal angles. The ratio of the emitted intensity to the square of the incident intensity, $I(2\omega)/I_0^2(\omega)$, is plotted. The angle of incidence, ϕ_0, is 67.5°. The incident and outgoing light are p-polarised [18].

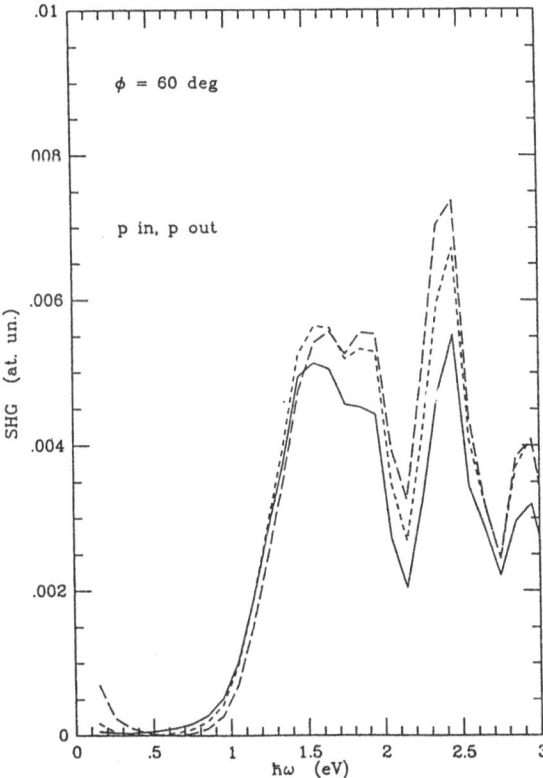

Fig. 7.8. Calculated SHG from Si(111) 1x1-As for various angles of incidence. The ratio of the emitted intensity to the square of the incident intensity, $I(2\omega)/I_0^2(\omega)$, is plotted. The azimuthal angle, ψ, is 60°. The incident and outgoing light are p-polarised [18].

7.3 Supercell Formalisms

An alternative approach which may be used to overcome the difficulties arising from the loss of periodicity in the z direction, perpendicular to the surface, is to consider a model bulk system obtained by a periodic repetition, in the z-direction, of a suitably chosen supercell of thickness d.

If the EM field is adiabatically switched on at $t = -\infty$, and the wavevector of the radiation is very small compared to the dimension of the BZ, one can adopt the following dipole-like approximations:

$$\mathbf{E}^{ex}(\mathbf{r}, t) \approx \mathbf{E}^{ex}(t) \tag{7.10}$$

$$\mathbf{E}(\mathbf{r},t) \approx \mathbf{E}^{ex}(t) + \mathbf{E}^{ind}(z,t) \tag{7.11}$$

where the perturbative electric field is thought of as a superposition of an external, approximately spatially-constant field, and an induced field, $\mathbf{E}^{ind}(z,t)$, which is a periodic function of z whose period is the thickness, d, of the supercell.

The Green function formalism is set out in Appendix C. Let us consider a model system described by the effective one-particle, Hermitian Hamiltonian, H (which contains, in general, a non-local potential). The Hamiltonian in the presence of a classical EM field can be obtained by the minimal coupling replacement in the unperturbed Hamiltonian. In the Coulomb gauge up to second order in the vector potential $\mathbf{A}(\mathbf{r},t)$, one obtains the following expression for the interaction Hamiltonian:

$$H_{int} = \frac{e}{c}\mathbf{A}\cdot\mathbf{v} + \frac{1}{2}\left(\frac{e}{c}\right)^2 \frac{1}{i\hbar}\mathbf{A}\cdot[\mathbf{r},\mathbf{A}\cdot\mathbf{v}] \tag{7.12}$$

where the velocity operator is defined by:

$$\mathbf{v} = \frac{1}{i\hbar}[\mathbf{r},H] \tag{7.13}$$

By extending the treatment of Sect. 2.2 to include higher order effects, we can obtain the nonlinear response to an optical perturbation. One obtains the following expression for the second-order, frequency-dependent, density operator:

$$\rho^{(2)}(\omega',\omega'') = -\frac{ie}{\omega'+\omega''}\sum_k \sum_{jj'j''}\left\{\left[\frac{\Delta f_{jj'}^k <\psi_{j'k}|\mathbf{E}^{(2)}(\omega',\omega'')|\psi_{j''k}>\cdot\mathbf{v}_{j''j}^k}{\hbar(\omega'+\omega'')-\Delta E_{j'j}^k}\right]\right.$$

$$-\frac{e^2}{\omega'\omega''}\frac{1+P_{12}}{2}\sum_{JJ'}\frac{<\psi_{j'k}|E_\gamma^{(1)}(\omega'')|\psi_{Jk}><\psi_{j''k}|E_\beta^{(1)}(\omega')|\psi_{J'k}>}{\hbar(\omega'+\omega'')-\Delta E_{j'j}^k}$$

$$\times\left[\frac{\Delta f_{jj'}^k}{2}\delta_{Jj''}W_{\gamma\beta J'j}^k + \left(\frac{\Delta f_{jj''}^k}{\hbar\omega'-\Delta E_{j''j}^k}-\frac{\Delta f_{j''j'}^k}{\hbar\omega''-\Delta E_{j'j''}^k}\right)v_{\gamma J Jj''}^k v_{\beta J'j}^k\right]\right\}$$

$$\times|\psi_{j'k}><\psi_{jk}| \tag{7.14}$$

In equation 7.14, and in what follows, α and β label the Cartesian components, and the Einstein sum convention over Greek indexes has been adopted; f_{jk} is the occupation number of the state jk, and $\psi_{jk}(t)$ is the time-evolved wave function. The following tensor operator has also been introduced:

$$W_{\alpha\beta} = \frac{1}{i\hbar}[r_\alpha, v_\beta] \tag{7.15}$$

and P_{12} interchanges the two photons, in particular ω' and ω'' in equation 7.14. It is worth noting that the above operator is invariant under translations.

Expression 7.14 is valid for systems described by Hamiltonians containing both non-local and local potentials. Moreover, if the density of direct current is negligible, one can obtain the following expressions for the bulk first and second order susceptibilities:

$$\chi^{(1)}_{\alpha\beta}(z;z',\omega') = -d\frac{e^2}{\omega'^2}\int\frac{d\mathbf{k}}{(2\pi)^3}\sum_{jj'JJ'}\frac{\Delta f^k_{jj'}}{\hbar\omega' - \Delta E^k_{j'j}}v^k_{\alpha Jj'}v^k_{\beta J'j}P^k_{z jJ}P^k_{z'j'J'} \tag{7.16}$$

$$\chi^{(2)}_{\alpha\beta\gamma}(z;z',\omega';z'',\omega'') = \frac{id}{\omega'+\omega''}\frac{e^3}{\omega'\omega''}\frac{1+P_{12}}{2}\int\frac{d\mathbf{k}}{(2\pi)^3}$$

$$\times\sum_{jj'j''}\sum_{JJ'J''}\frac{v^k_{\alpha Jj'}}{\hbar(\omega'+\omega'')-\Delta E^k_{j'j}}P^k_{z jJ}P^k_{z'j'J'}P^k_{z''j''J''}$$

$$\times\left[\frac{\Delta f^k_{jj'}}{2}\delta_{J'j''}W^k_{\gamma\beta J''j} + \left(\frac{\Delta f^k_{jj''}}{\hbar\omega'-\Delta E^k_{j''j}} - \frac{\Delta f^k_{j''j'}}{\hbar\omega''-\Delta E^k_{j'j''}}\right)v^k_{\gamma J'j''}v^k_{\beta J''j}\right] \tag{7.17}$$

In equation 7.17, P_{12} interchanges ω' and ω'', β with γ, and z' with z''. In the above equations, matrix elements of operators of the form:

$$\mathbf{K} \equiv [\mathbf{r}, \mathbf{A}] \tag{7.18}$$

appear, where $\mathbf{A}$ is an operator that commutes with the translation operator, $\mathbf{T_R}$, defined by:

$$\mathbf{T_R}f(\mathbf{r}) = f(\mathbf{r}+\mathbf{R}) \tag{7.19}$$

for each Bravais lattice vector, $\mathbf{R}$. It is worth noting also that $\mathbf{K}$ commutes with $\mathbf{T_R}$. However, the operator $\mathbf{r}$ does not have this property, so that the product $\mathbf{r}\psi_{jk}(\mathbf{r})$ does not transform under the action of $\mathbf{T_R}$ as a Bloch function. As a consequence, the matrix elements of $\mathbf{K}$ cannot be calculated by using the turn-over rule:

$$\mathbf{K}_{j'j}^{k} \neq \sum_{j''}\left[\mathbf{r}_{j'j''}^{k} A_{j''j}^{k} - A_{j'j''}^{k}\mathbf{r}_{j''j}^{k}\right]\tag{7.20}$$

Moreover, while the matrix elements of the operator $\mathbf{K}$ are independent of the choice of the primitive cell, the matrix elements of $\mathbf{r}$ are, in contrast, not well-defined in value. Nevertheless, by introducing a vectorial operator, λ, defined as:

$$\lambda \psi_{jk}(\mathbf{r}) \equiv -i\frac{\partial}{\partial \mathbf{k}}\psi_{jk}(\mathbf{r})\tag{7.21}$$

one can obtain:

$$\mathbf{K}_{j'j}^{k} = \sum_{j''}\left[\tilde{\mathbf{r}}_{j'j''}^{k} A_{j''j}^{k} - A_{j'j''}^{k}\tilde{\mathbf{r}}_{j''j}^{k}\right] + i\frac{\partial}{\partial \mathbf{k}}A_{j'j}^{k}\tag{7.22}$$

where

$$\tilde{\mathbf{r}} = \mathbf{r} - \lambda\tag{7.23}$$

Then one has:

$$v_{j'j}^{k} = \frac{i}{\hbar}\Delta E_{j'j}^{k}\tilde{r}_{j'j}^{k} + \frac{1}{\hbar}\frac{\partial}{\partial \mathbf{k}}E_{jk}\delta_{j'j}\tag{7.24}$$

and

$$W_{\beta\gamma j'j}^{k} = \frac{1}{\hbar^{2}}\sum_{j''}\left[\tilde{r}_{\beta j'j''}^{k}\tilde{r}_{\gamma j''j}^{k} + \tilde{r}_{\gamma j'j''}^{k}\tilde{r}_{\beta j''j}^{k}\right]\Delta E_{j''j}^{k}$$
$$-\frac{1}{\hbar^{2}}\Delta E_{j'j}^{k}\sum_{j''}\tilde{r}_{\gamma j'j''}^{k}\tilde{r}_{\beta j''j}^{k} + \frac{i}{\hbar}\left[\left(v_{\gamma j'j'}^{k} - v_{\gamma jj}^{k}\right)\tilde{r}_{\beta j'j}^{k} + \left(v_{\beta j'j'}^{k} - v_{\beta jj}^{k}\right)\tilde{r}_{\gamma j'j}^{k}\right]$$
$$+\frac{1}{\hbar^{2}}\Delta E_{j'j}^{k}\frac{\partial}{\partial k_{\beta}}\tilde{r}_{\gamma j'j}^{k} + \frac{1}{\hbar}\frac{\partial}{\partial k_{\beta}}v_{\gamma jk}\delta_{j'j}$$

$$\tag{7.25}$$

Finally, it is worth noting that if one writes the Bloch function, ψ_{jk}, in the form:

$$\psi_{jk}(\mathbf{r}) = \exp(i\mathbf{k}\cdot\mathbf{r})u_{jk}(\mathbf{r})\tag{7.26}$$

one has:

$$\tilde{\mathbf{r}}^k_{j'j} = \int\limits_{cell} dr' u^*_{j'k}(r') i \frac{\partial}{\partial \mathbf{k}} u_{jk}(r') \tag{7.27}$$

so that the $\tilde{\mathbf{r}}$ operator is also Hermitian, due to the orthonormality relation.

7.4 Other Techniques

The study of nonlinear optical coefficients of surfaces and interfaces has highlighted a problem already implicit in previous work on the corresponding bulk properties. If the unit cell is large, so that many basis functions are used in expressing the electronic states, the usual formulae for nonlinear coefficients [20] entail the summation of huge numbers of terms, increasing exponentially with the order, n, of the optical effect considered.

While there have been several more-or-less successful calculations of $\chi^{(2)}$ of bulk semiconductors [21, 22], problems involving more than two atoms per unit cell have hardly begun to be tackled [23]. Only limited calculations of $\chi^{(3)}$ have been attempted [24].

The slab formulation for surface properties necessarily presents us with this kind of difficulty, even if the 2-dimensional unit cell is quite small, because of the large number of layers needed. Thus the calculations undertaken by the group of Del Sole, based on the Cini formalism, which are reviewed in Sect. 7.2 above, break new ground in the number of basis states used. Although founded on a more complete theory of SHG, they pose a computational challenge similar to that of the earlier work on bulk solids which has been cited.

In Sect. 2.3, we have outlined one approach to the circumvention of this computational bottleneck, namely a retreat to the localised description in terms of the response of individual bonds. It remains confined to the static ($\omega \rightarrow 0$) limit, at least for now.

In search of another approach, inspiration was sought in the various attempts made over the last twenty years to calculate electronic properties in the solid state without recourse to diagonalisation of the Hamiltonian and the performance of explicit sums of the eigenstates in the resulting band structure. The most popular of these alternative approaches has perhaps been the recursion method [25], but this is not easily extended to optical properties. A somewhat analogous technique, the equation-of-motion method [26], has proved to be much more adaptable : it has been applied to localisation [27], conductivity [28] and the Hall effect [29], using both pseudopotential and tight-binding Hamiltonians.

Accordingly Weaire, Hickey and Morgan [30] have begun to develop appropriate formulations for linear optical properties. These depend on relating band summations or averages to appropriate time integrals of matrix elements between time-dependent random wave functions. As in the previous applications of the equation-of motion method, such algorithms are well adapted to large systems, in which the fluctuations due to the random character of the wave functions are relatively small. For large systems the method is economical in time and storage, and both increase only linearly with the order of the perturbation term being evaluated.

For example, the dielectric function , $\varepsilon(\omega)$, is represented by

$$\varepsilon(\omega) = \frac{G\hbar^2}{\pi E^2} \text{Im} \left[\lim_{T \to \infty} T^{-1} \frac{i}{\hbar} \int_0^T dt \, <c(t)| \, p| v(t)> \exp(-i[E + i\eta]t/\hbar) \right.$$
$$\left. \times \int_t^\infty dt' \exp(i[E + i\eta]t'/\hbar) <v(t')| \, p| c(t')> \right] \tag{7.28}$$

Here G is $2\Omega^{-1}(2\pi e/m)^2$, $E = \hbar\omega$, v and c are random wavefunctions confined to the valence and conduction bands, η is an infinitesimal parameter (or a finite broadening parameter), and p denotes the appropriate component of the momentum operator.

This is well adapted to straightforward computation, which consists essentially of the time-integration of Schrödinger's Equation for each random wave function, the evaluation of matrix elements of **p** and the updating of the time integrals. This is especially efficient in any simple tight-binding scheme for which the operation of the Hamiltonian, H, or the operator, **p**, is particularly compact, non-zero elements being confined to short range.

First results from this approach, for a 216-atom model of amorphous Si, have been encouraging [30, 31]. Corresponding expressions have been derived for second and third order nonlinear optical properties [32]. These expressions have been used for model calculations for the second and third order optical spectra of bulk GaAs. The third harmonic spectra for crystalline and amorphous silicon are also being compared. If this method fulfils its apparent promise, it may provide an expeditious style of calculation for large structural models, most suitable whenever broad features of the response as a function of frequency are at issue.

A recent publication [33] is aimed in a similar direction, but uses time-integration of the Schrödinger Equation in a spirit of direct summation, incorporating an applied AC field, while we have set out to use it as a method of evaluation of standard perturbation expressions.

References

1.   See, e.g., Bloembergen, N.: Nonlinear Optics. Benjamin, New York 1965. *Ibidem* in: Nonlinear Spectroscopy, Proceedings of the International School of Physics, Enrico Fermi, Course LXIV. Shen, Y.R., Bloembergen, N. (eds.)North-Holland, Amsterdam, 1977

2. Shen, Y.R.: The Principles of Nonlinear Optics. Wiley, New York 1984.
3. Cini, M.: Physical Review *B 43*, 4792 (1991)
4. Cini, M., Del Sole, R., Ping, J.G., Reining, L.: Proceedings of the Second Epioptic Workshop, Berlin 1991, p. 1
5. Cini, M.: Physical Review *B 17*, 2486 (1978). *Ibidem*: Surface Science *79*, 589 (1979)
6. Schuda, F., Stroud, C.R., Hercher, M.: Journal of Physics B: Atomic and Molecular Physics *7*, L198 (1974)
7. Haken, H.: Light, Volume 1, Waves, Phtons and Atoms. North-Holland, Amsterdam, 1981
8. Carmichael, H.J., Walls, D.F.: Journal of Physics B: Atomic and Molecular Physics *9*, 1199 (1976)
9. Cini, M., D'Andrea A., Verdozzi, C.: Physics Letters A *180*, 430 (1993). *Ibidem*, Inernational Journal of Modern Physics, in press (1995)
10. Cini, M.: Surface Review and Letters *1*, 443 (1994)
11. Selloni, A., Marsella P., Del Sole, R.: Physical Review *B 33*, 8885 (1986)
12. Patterson, C.H., Messmer, R.P.: Physical Review *B 39*, 1372 (1989)
13. Kelly, P.V., Tang, Z.-R., Woolf, D.A., Williams, R.H., McGilp, J.F.: Surface Science *251/252*, 87 (1991)
14. Hybertsen, M.S., Louie, S.G.: Physical Review *B 38*, 4033 (1988)
15. Vogl, P., Hjalmarson, H.P., Dow, J.D.: Journal of the Physics and Chemistry of Solids *44*, 365 (1983)
16. Manghi, F., Del Sole, R., Selloni, A., Molinari, E.: Physical Review *B 41*, 9935 (1990)
17. Scandolo, S., Bassani, F.: Physical Review *B 44*, 8466 (1991)
18. Reining, L., Del Sole, R., Cini, M., Ping, J.G.: Physical Review *B 50*, 8411 (1994)
19. McIntyre, J.D.E., Aspnes, D.E.: Surface Science *24*, 417 (1971)
20. Butcher, P.N., Cotter, D.: The Elements of Nonlinear Optics. Cambridge University Press, Cambridge 1990
21 Fong, C.Y., Shen, Y.R.: Physical Review *B 12*, 2325 (1975)
22. Moss, D.J., Sipe, J.E., van Driel, H.M.: Physical Review *B 36*, 9708 (1987)
23 Ghahramani, E.D., Moss, D.J., Sipe, J.E.: Physical Review Letters *64*, 2815 (1990)
24. Moss, D.J., Ghahramani, E., Sipe, J.E., van Driel, H.M.: Physical Review *B 41*, 1542 (1990)
25. Haydock, R., Heine, V., Kelly, M.J.: J.ournal of Physics C: Condensed Matter *5*, 2845 (1992)
26. Alben, R., Blume, M., Krakauer, H., Schwartz, L.: Physical Review *B 12*, 4090 (1975)
27. Weaire, D., Williams, A.R.: Journal of Physics *C 10*, 1239 (1977)
28. Hickey, B.J., Burr, J.N., Morgan, G.J.: Philosophical Magazine Letters *61*, 161 (1990)
29. Holender, J.M., Morgan, G.J.: Philosophical Magazine Letters *65*, 225 (1992)
30. Weaire, D., Hickey, B.J., Morgan, G.J.: Journal of Physics: Condensed Matter *3*, 9575 (1991)
31. Weaire, D., Hobbs, D., Morgan, G.J., Holender, J.M., Wooten, F.: Journal of Non-Crystalline Solids *164/166*, 877 (1993)
32. Hobbs, D.: PhD dissertation, University of Dublin , Dublin 1995
33. Plaja, L., Roso-Franco, L.: Physical Review *B 45*, 8334 (1992)

Chapter 8. Second Harmonic and Sum Frequency Generation

John McGilp

Department of Physics, Trinity College, Dublin 2, Ireland

8.1 Phenomenology

As mentioned in Sect. 1.5, the lowest order nonlinear optical response of materials produces three-wave mixing phenomena, which include SHG and SFG, and these phenomena may be surface sensitive at non-destructive power densities. For centrosymmetric materials, an order of magnitude calculation shows that the surface effect should be at least comparable in size to the higher order nonlocal bulk effects [1]. Initial surface SHG studies in the 1960s under non-UHV conditions detected a surface signal but found no dependence on adsorbate or surface structure, and this held back the development of the field until the early 1980s, when Shen in Berkeley established the potential of SHG as a surface probe [2].

The first UHV study, where conventional surface probes were used to characterise the surface, appeared in 1984 [3]. It was shown that the adsorption of O and CO damped the SHG signal from Rh(111), while the adsorption of Na enhanced it. The following year Heinz and co-workers showed that the azimuthal dependence of the SHG signal was sensitive to the symmetry change between the (2x1) and the (7x7) reconstructions of the Si(111) surface [4]. Later, McGilp and Yeh used the Si(111)-Au system to show that SHG could provide structural information about the buried metal-semiconductor interface [5, 6]. As has been emphasised previously, epioptic techniques have to be tested on well-characterised surfaces and interfaces, and SHG studies of this type were important in providing good evidence of surface and interface sensitivity at the monolayer level. A number of reviews of SHG applied to thin films, surfaces and interfaces have been published [7-11]. Richmond *et al* produced a comprehensive review of the field in 1988 [8] and, more recently, Heinz has discussed the phenomenology of surface and bulk effects in centrosymmetric media [11].

8.1.1 Theory of the Nonlinear Optical Response

Calculation of the surface nonlinear optical response from first principles is difficult, as discussed in previous chapter. The phenomenology, however, is well known and allows symmetry arguments to be used in the interpretation of experimental data [4, 5, 12, 13]. In the most general second order nonlinear response of a system, three-wave mixing occurs in which two incident fields of frequency ω_1 and ω_2 combine to produce a third field of frequency ω_3, where

$$\omega_3 = \omega_1 \pm \omega_2 \tag{8.1}$$

The four possible combinations and degeneracies in equation 8.1 can produce SHG, optical rectification, SFG and difference frequency generation [14]. It is important to note that these processes are coherent and the radiating fields have a well-defined direction. For example, SHG from a surface in the usual reflection geometry in vacuum emerges along the path of the reflected primary beam, which simplifies detection dramatically. This coherent property enables SHG and SFG to be used as surface probes, even though the cross-section for the process is typically four orders of magnitude smaller than the RS cross-section (Sect. 5.2).

The phenomenology of bulk and surface SHG and SFG from cubic centrosymmetric crystals will only be outlined here, as it is discussed in detail elsewhere [11, 13]. In the bulk of centrosymmetric crystals the electric dipole term is zero, and the lowest order nonlinear polarisation density is of magnetic dipole and electric quadrupole symmetry [15]. For SHG, the effective polarisation can be written as [11, 13]:

$$P_i(2\omega;\mathbf{r}) = \varepsilon_0 \left\{ \chi_{ijkl}^{NL} E_j(2\omega;\mathbf{r}) \nabla_k (2\omega;\mathbf{r}) E_l(2\omega;\mathbf{r}) \right\} \tag{8.2}$$

where χ_{ijkl}^{NL} is the fourth-rank tensor component describing the nonlocal response, and the gradient is determined with respect to the field coordinates. For cubic crystals this expression reduces to:

$$P_i(2\omega) = \varepsilon_0 \left\{ (\delta - \beta - 2\gamma)(\mathbf{E} \cdot \nabla) + \beta(\nabla \cdot \mathbf{E})E_i + \gamma \nabla_i(\mathbf{E} \cdot \mathbf{E}) + \zeta E_i \nabla_i E_i \right\} \tag{8.3}$$

where δ, β, γ are isotropic parameters, and ζ is an anisotropic parameter, all simply related to the few non-zero components of χ_{ijkl}^{NL} [11, 13]. Only the last two terms contribute if excitation is by a single transverse plane wave. Surface SHG can arise from the higher order terms of equation 8.3, due to the large field gradients normal to the surface, but it can also arise from an electric dipole term because the inversion symmetry is now broken at the surface [1, 12]. The electric dipole term is of the form:

$$P_i(2\omega) = \varepsilon_0 \{ \chi_{ijk}^s E_j(\omega) E_k(\omega) \} \qquad (1.17)$$

where χ_{ijk}^s is the second-order susceptibility tensor component reflecting the structure and symmetry properties of the surface or interface, as discussed in Sect. 1.5. Expressions for the SH polarisation can be obtained by applying boundary conditions [16], or by a more general Green function approach [17]. Sipe *et al* [13] have tabulated expressions for the total SH fields from the (001), (110) and (111) faces of cubic centrosymmetric crystals. Where surface and bulk contributions are of similar size it can be difficult to distinguish the two contributions on the basis of symmetry arguments alone [12, 18], and this caused some controversy in earlier work [19-21]. However, there are now many examples, particularly from semiconductors, where SHG from a surface or interface has been clearly identified [8-11].

SFG, where the two incident fields are different, has lower symmetry than SHG. Table 8.1 lists the independent non-vanishing tensor elements for SHG and SFG from surfaces and interfaces of different symmetries, from which structural information can be deduced.

Experiments at normal incidence, for example, provide in-plane symmetry information about the surface [4, 23] or interface [5, 6]. For a crystalline solid with a surface or interface in the xy-plane, and normally-incident radiation linearly polarised at an angle, φ, to the x-axis of the crystal, the SHG intensity polarised along the x- and y-axes is given by [6]:

$$I_x^{2\omega}(\varphi) \sim \left| \chi_{xxx}^s \cos^2 \varphi + \chi_{xyy}^s \sin^2 \varphi + \chi_{xyx}^s \sin 2\varphi \right|^2 \qquad (8.4)$$

$$I_y^{2\omega}(\varphi) \sim \left| \chi_{yxx}^s \cos^2 \varphi + \chi_{yyy}^s \sin^2 \varphi + \chi_{yxy}^s \sin 2\varphi \right|^2 \qquad (8.5)$$

where χ_{ijk}^s is the surface tensor component. The presence of symmetry elements simplifies these expressions. For example, Table 8.1 shows that components xyx, yyy, yxx are zero for 1m symmetry, while all in-plane components are zero for 2mm and 4mm symmetries, with no SHG being detected using this experimental geometry.

Off-normal excitation brings the z components into play, and careful choice of experimental geometry and polarisation vectors is often important, particularly with lower symmetry interfaces, where there may be many non-zero components. In deriving appropriate analytic expressions, it is necessary to transform quantities

Table 8.1. Independent non-zero elements of χ_{ijk}^s for crystallographic and continuous point groups for a surface in the xy-plane [22]. For SHG, $\chi_{ijk}^s = \chi_{ikj}^s$ but, for SFG, these elements may be distinct and any symmetry are considerations indicated in parentheses. Terms entirely in parentheses are only present for SFG (after [11]).

Symmetry class	Independent non-zero elements of χ_{ijk}^s
1	xxx, xxy, xyy, yxx, yxy, yyy, xxz, xyz, yxz, yyz, zxx, zxy, zyy, xzz, yzz, zxz, zyz, zzz
1m †	xxx, xyy, xzz, xzx, yzy, yxy, zxx, zyy, zxz, zzz
2	xzx, xyz, yxz, yzy, zxx, zyy, zxy, zzz
2mm	xzx, yzy, zxx, zyy, zzz
3	$xxx = -xyy = -yyx\ (= -yxy)$, $yyy = -yxx = -xyx\ (= -xxy)$, $yzy = xzx$, $zxx = zyy$, $xyz = -yxz$, zzz, $(zxy = -zyx)$
3m †	$xxx = -xyy = -yxy\ (= -yyx)$, $xzx = yzy$, $zxx = zyy$, zzz
4, 6, ∞	$xxz = yyz$, $zxx = zyy$, $xyz = -yxz$, zzz, $(zxy = -zyx)$
4mm, 6mm, ∞m	$xxz = yyz$, $zxx = zyy$, zzz

† One mirror plane is perpendicular to $\hat{y}$

in the laboratory axes into coordinates associated with the principal axes of the crystal [13, 24, 25]. Experiments under ambient conditions typically involve sample rotation plots, where the m-polarised SH intensity, $I_{mn}^{2\omega}(\psi)$, for an n-polarised pump beam, is measured as a function of ψ, the azimuthal angle between the ξ-axis of the crystal and the plane of incidence (the laboratory xz-plane) (Fig. 8.1). Polarisations of high symmetry with respect to the optical plane, particularly s- and p-polarisations, are used. Explicit analytic expressions for this type of measurement on (001), (111) and (110) faces of cubic crystals have been published in a convenient form by Sipe $et\ al$ [13]. For example, with a (111) surface of 3m symmetry, such as Si(111)7x7, the surface SH intensities are given by:

$$I_{ss}^{2\omega}(\psi) \sim \left| f_1 \chi_{xxx}^s \sin(3\psi) E_s(\omega)^2 \right|^2 \tag{8.6}$$

$$I_{sp}^{2\omega}(\psi) \sim \left| f_2 \chi_{xxx}^s \sin(3\psi) E_p(\omega)^2 \right|^2 \tag{8.7}$$

$$I_{ps}^{2\omega}(\psi) \sim \left| [f_3 \chi_{zxx}^s + f_4 \chi_{xxx}^s \cos(3\psi)] E_s(\omega)^2 \right|^2 \tag{8.8}$$

$$I_{pp}^{2\omega}(\psi) \sim \left| \{[f_5 \chi_{zzz}^s + f_6 \chi_{zxx}^s + f_7 \chi_{xzx}^s] + f_8 \chi_{xxx}^s \cos(3\psi)\} E_p(\omega)^2 \right|^2 \tag{8.9}$$

where f_i are Fresnel factors, the x-axis is parallel to <112>, and the z-axis is along the surface normal.

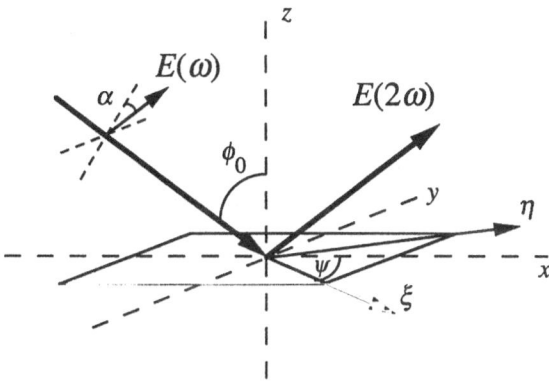

Fig. 8.1. Reflection SHG geometry. The fundamental is plane-polarised at an angle α to the plane of incidence (see also Fig. 1.2).

The role of the Fresnel factors has been reviewed by Heinz [11]. Briefly, two-phase and three-phase models, analogous to those describing the linear response in Sect. 1.3, have been used to determine the EM field amplitudes. However, the rapid variation of field amplitudes in the selvedge region has already been mentioned at the beginning of Sect. 2.1 and, given that the nonlinear polarisability depends on the square of the amplitude of the exciting field, greater sensitivity to this variation may be expected. It is not yet clear whether the models which are being used, quite successfully, to describe the linear optical response, are adequate for the nonlinear response. The effect of local fields on the SH response, also discussed in Sect. 2.1, provides an additional complication. Quantitative measurements of this effect on well-characterised liquid crystal monolayers have been reported [26].

In general, both the susceptibility components and the Fresnel factors may be complex. The interpretation of the overall SH response may also be complicated by contributions from different tensor components, as is apparent from equations 8.8 and 8.9. This is particularly so for the commonly-used pp-configuration. However, equations 8.6 and 8.7 show that χ^s_{xxx} can be measured independently, and equation 8.8 shows that, by choosing $\psi = 30°$ and an s-input/p-output configuration, the χ^s_{zxx} component can also be measured independently. This type of approach allows a simpler comparison of experiment and microscopic theory, and is particularly useful for coverage-dependent studies [27-32].

Sample rotation studies are difficult in UHV systems. An alternative approach can be used which involves rotating the input polarisation angle, instead of the sample. The plane of incidence is typically aligned along an azimuth corresponding to a symmetry line in the surface plane, and the experiment is repeated for other, inequivalent, symmetry lines [33].

The phase of the (complex) SH signal provides important extra information, particularly where a full spectroscopic study (Sect. 8.5) is not possible. Measurements of the phase of the SH signal from surfaces and thin films are rarely reported, although methods of measuring the SH phase by interferometry are well-known [34], and recently have been used under UHV conditions for the first time [29]. The extra information available concerns the presence of nearby electronic resonances, at either the excitation frequency or the SH frequency. Within the dipole approximation, the second-order perturbation theory expression for χ^s_{ijk} has terms of the form [1]:

$$\chi^s_{ijk} \sim \sum_{n,n'\neq g} \frac{\langle g|i|n\rangle}{(2\omega - \omega_{ng} + i\Gamma_{ng})} \frac{\langle n|j|n'\rangle\langle n'|k|g\rangle + \langle n|k|n'\rangle\langle n'|j|g\rangle}{(\omega - \omega_{n'g} + i\Gamma_{n'g})} \qquad (8.10)$$

where $\langle g|i|n\rangle$ is the dipole moment transition matrix element along the i-coordinate between the electronic ground state of the interface, g, and an excited state, n, with Γ^{-1}_{gn} being a characteristic relaxation time. In equation 8.10, the denominator becomes small near either an ω or 2ω resonance, enhancing the SH intensity. Resonances provide an important spectroscopic dimension to SHG from surfaces and interfaces, but it is difficult to scan over wide frequency ranges looking for resonance phenomena, and little spectroscopic work from well-characterised semiconductor interfaces has been reported so far [33, 35-38]. Phase measurements, in contrast, are quite simple (see below) although, of course, they provide less complete information. The phase angle, ϕ, of the surface tensor component, in $\chi^s_{ijk} = |\chi^s_{ijk}|e^{i\phi}$, can be related to the phase of the SH intensity, measured interferometrically. Off-resonance, $\phi = 0°$ or $180°$, and χ^s_{ijk} is real, with sign ± 1. For $\phi = 90°$ or $270°$, χ^s_{ijk} is exactly on resonance and is purely imaginary. For intermediate values of the phase, χ^s_{ijk} is complex, and there are resonances nearby [28, 29]. However, the phase shift depends on both the energy and the width of the resonance, and at least two phase measurements at different wavelengths near resonance have to be made to allow these parameters to be determined quantitatively.

8.2 Experiment

Pulsed laser systems, with wavelengths lying between 500 nm and 2500 nm, are the main excitation source for three-wave mixing experiments in semiconductors. CW laser excitation of surface SHG has been reported [39], but surface damage considerations, together with the advantages of simple gating electronics, favour pulsed sources. A typical experiment might use nanosecond pulses of a few mJ energy, into a beam diameter of a few mm and with an energy density kept below 1 kJ m^{-2} to avoid any laser-induced desorption or damage effects. Q-switched Nd:YAG lasers have been widely used at 1064 nm excitation and, frequency-doubled, at 532 nm, while dye, Ti:sapphire and optical parametric oscillator (OPO) systems [40, 41] are beginning to be used for wavelength-dependent studies [29, 30, 32, 33, 35-38, 42].

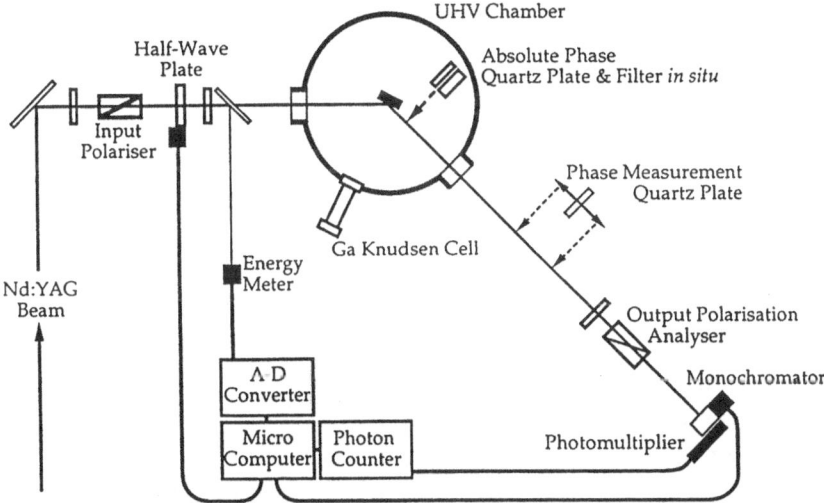

Fig. 8.2. Schematic of an oblique-incidence UHV experiment for measuring both the intensity and phase of a SH signal (after [29]).

A typical UHV experimental configuration for SHG at oblique incidence is shown in Fig. 8.2. Glan-Taylor prisms, together with a half-wave-plate or double Fresnel rhomb are used for polarisation selection. It is very important to remove the SHG signal generated in these optical components in the input line with optical filters. The SHG signal is small, and either a gated integrator or a gated photon counter is used The energy meter can be replaced by a second SHG line which uses a bulk SHG signal, for example from urea, to provide pulse-by-pulse normalisation of the SHG signal. Such systems have a remarkable discriminating power of 10^{18} between primary and SH photons. The SH intensity can be calibrated by inserting an x-cut quartz plate in the input beam and observing the Maker fringes [43] produced by SHG in the bulk of the quartz. This calibration technique has the advantage that the experimental geometry, and hence overall system sensitivity, remains unchanged.

The phase of the SH signal can be measured by nonlinear interferometry. A quartz plate is inserted in the output line and traversed along the beam, while the combined SH intensity from the sample and the quartz is monitored. The dispersion in air of the fundamental and SH signals produces a variation in the optical path length of the two SH signals, allowing their phase difference to be measured from the interference signal obtained [1]. Varying the pressure in a gas cell can also be used to change the optical path length [34]. If a second quartz plate is rotated into the output beam, within the UHV chamber and without disturbing the optical path then, by comparing the phase difference between the two quartz plates, the phase shift arising from the UHV window can be eliminated [29], and the absolute phase of the system measured using the quartz reference which, in the frequency range of interest here, has an absolute phase angle of zero for $\chi^{(2)}_{xxx}$ [44].

8.3 Clean Si Surfaces

A number of *in situ* UHV studies have now been performed on clean Si surfaces under UHV conditions. Strong SH signals are observed from Si(111)7x7, Si(111)2x1 and Si(100)2x1 surfaces using 1064nm excitation, with the surface dipolar response being much larger than the higher order bulk response. This is a difficult region of the spectrum for pulsed lasers, and phase measurements have proved to be of particular value here.

As discussed in Sect. 8.1, azimuthal studies of rotational anisotropy at, or near, normal incidence can be used to probe in-plane symmetry. This approach works particularly well for the (111) surface of elemental semiconductors. The first such study of well-characterised surfaces was reported by Heinz *et al* [4, 45]. The SH signal, using 1064 nm excitation, from Si(111)2x1 and Si(111)7x7 surfaces is shown in Fig. 8.3, and is clearly sensitive to the symmetry change between the two reconstructions, the solid lines being fits to equations 8.4 and 8.5. This work was important because it showed, in a particularly clear way, that surface information was available from SHG [4, 45].

Spectroscopic studies in this region have only recently become possible with the development of reliable OPO systems, but McGilp, Rasing and co-workers made the first absolute phase measurements in UHV to show that the Si(111)7x7 surface has a nearby resonance, for 1064 nm (1.17 eV) excitation [29]. No enhancement is seen for 532 nm excitation, indicating that the resonance is associated with the fundamental frequency, at 1064 nm, rather than the SH frequency. In a later study, McGilp and Power [30] used the adsorption of Sb atoms at very low coverages to show that the resonance was due electronic states associated with the Si adatoms of the dimer-adatom-stacking fault (DAS) structure [46]. STM work had shown that, on adsorption, Sb adatoms simply replace the Si adatoms at coverages around 0.03 ML [47]. The first calculations, by Del Sole and co-workers [47], of the *linear* optical properties of the Si(111)7x7 surface have just appeared, and provide further evidence for this, in that transitions between the filled adatom back bond and the empty adatom dangling bond states occur around 1.4 eV. In the SHG study, both the χ_{zxx}^s and χ_{xxx}^s components were shown to be sensitive to displacement of the Si adatoms [30], and the involvement of backbond states, which have a non-zero component in the surface plane, would nicely account for this behaviour. Spectroscopic SHG studies of the Si(111)7x7 surface are currently underway [49]. The temperature dependence of the SH signal in the region of the Si(111)7x7 → 1x1 order-disorder phase transition is also being studied, and shows an abrupt intensity change [50, 51].

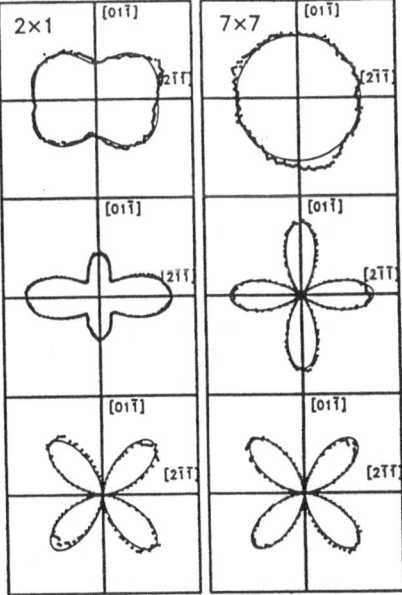

Fig. 8.3. SH intensity from Si(111)2x1 and Si(111)7x7 surfaces, as a function of the polarisation of the normally incident pump beam. Top: total SH intensity. Middle: SH intensity polarised along <112> Bottom: SH intensity polarised along <110> [4].

Less work has been reported on clean Si(001)2x1. Hollering and co-workers showed that the SH signal from this surface is also enhanced, relative to the bulk contribution, for 1064 nm excitation [27]. Enhancement was also observed by contrasting 634 nm with 1064 nm excitation [32, 42]. Phase measurements show a nearby resonance [52], and preliminary studies of the temperature dependence also support this conclusion [53].

A number of studies of clean, vicinal surfaces of Si(111) and Si(001) have been reported [5, 6, 33, 54, 55], showing the sensitivity of the SH response to step structure (see also Sect. 8.5).

8.4 Elemental Semiconductor-Adsorbate Systems

8.4.1 The Si(111)-Au System

The Si(111)-Au system (Fig. 8.4) was used for the first study of buried metal-semiconductor interfaces by SHG [5, 6]. The normal incidence study was able to conclude, by comparing results from Si(111) and Si(001) interfaces, together with LEED studies showing a disordered surface, that the patterns of Fig. 8.4 originated from ordered structures at the buried interface between the vicinal Si(111) substrate and the gold layer. The importance of spectroscopic SHG studies was also stressed in this work [6]. Off-normal studies of this system, which provide additional information, have recently been reported [56]. After room-temperature deposition, annealing produces Si(111)5x2-Au, Si(111)$\sqrt{3}$x$\sqrt{3}$-Au and Si(111)6x6-Au structures, as coverage increases to 1.2 ML [5, 6, 28]. The SH response in this coverage régime has been followed and shows a distinct maximum of the χ_{zxx}^{s} response on completion of the (5x2) structure, at 0.5 ML, for 1064 nm excitation [28]. SE results from the same system show a much smaller, but reproducible, change in the effective dielectric function at this coverage [31]. The advantage of SHG, where symmetry suppresses the bulk signal, over linear optical techniques, where small changes in a bulk-dominated response are detected, is clearly seen in this system.

8.4.2 Other Elemental Semiconductor-Metal Systems

In one of the earliest *in situ* studies, an order-of-magnitude increase in SH intensity on depositing 1 ML of Na on Ge(001) and Ge(111) was reported [57]. Barium deposition on Si(001)2x1 has also been studied, where a similar increase was observed [27]. By 4 ML Ba coverage, the out-of-plane components have increased by an order-of-magnitude, but the in-plane components are unchanged. The results were interpreted using a free-electron, Drude-like, bulk Ba model.

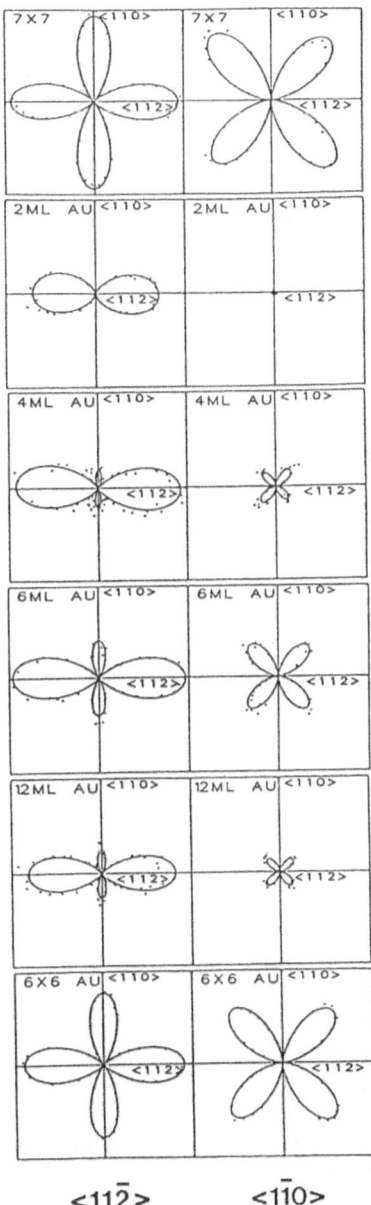

Fig. 8.4. SH intensity, polarised along the <112> and <110> azimuths, from Si(111)7x7 and Si(111)-Au [5].

This is a good example of the detailed information available when individual tensor components are identified by suitable choice of polarisation and azimuth [27].

SHG from Si(111)1x1-As and Si(001)2x1-As at four different wavelengths revealed resonance behaviour, and the first absolute values of the SH response from well-characterised surfaces in UHV were reported [42]. The Si(111)-Ga system has also been studied. Figure 8.5 shows the intensity variation and the phase shift, relative to the Si(111)7x7 value, for the χ^s_{zxx} component as a function of Ga coverage, using 1064 nm excitation. The variation in phase indicates the presence of an electronic resonance. The phase measurements were placed on an absolute scale by using a quartz reference in UHV, as discussed above [29]. The values obtained provided the first direct evidence that the SHG response from Si(111)7x7 is near resonance, for 1064nm excitation. The Si(111)√3x√3-Ga structure, which is complete where the maximum occurs in Fig. 8.5, was shown to be even closer to resonance. This contrasts with the off-resonance behaviour of the system, when using 634 nm excitation [58]. The structure has Ga adatoms in T4 sites [59-61], saturating the surface and removing all the dangling bond states [62]. The electronic structure of Si(111)√3x√3-Ga, determined by angle-resolved photoemission and inverse photoemission [63], reveals that suitable energy levels are accessible.

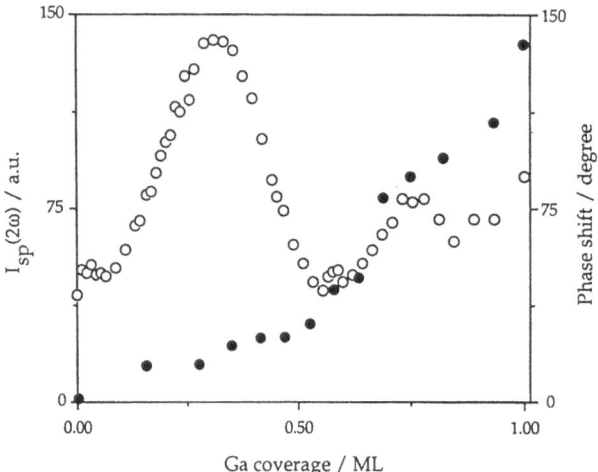

Fig. 8.5. SH intensity (∘) and phase shift (•) for s-polarised input and p-polarised output, as a function of Ga coverage, for 1064 nm excitation at $\psi = 30°$ to the <112> azimuth (after [29]).

The effect of Sb deposition on singular Si(111)7x7 and Si(001)2x1 has been studied up to 1 ML coverage [30, 32]. The use of low coverages of Sb on Si(111) to identify adatom resonances was discussed above. At higher coverages, very different behaviour is observed between the (001) and (111) surfaces, and between 1064 nm and 634 nm excitation. For the initial Si(001)2x1 surface, the results provide evidence of resonance enhancement when 1064 nm excitation is used [32]. This has now been confirmed by phase measurements [52].

8.4.3 Elemental Semiconductor-Gaseous Adsorbate Systems

The desorption kinetics and surface diffusion of hydrogen on singular Si(111)7x7 have been investigated using SHG [64, 65]. In the former case, isothermal desorption at low coverages was studied by using the high sensitivity of the SH response of Si(111)7x7 to adsorbates when using near-resonance 1064 nm excitation (Sect. 8.3). The desorption rate was found to deviate appreciably from the second-order behaviour expected for a simple associative desorption reaction, and the existence of different hydrogen adsorption sites on Si(111)7x7 was proposed [64]. In the second study, the diffraction of the SH signal from a submonolayer grating of adsorbed hydrogen, produced by laser-induced desorption, was followed as a function of temperature. This technique was first demonstrated by Shen and co-workers for the diffusion of CO on Ni(111) [66]. The advantage of this technique is its high sensitivity, with surface diffusivities < 10^{-14} cm^{2}s^{-1} being measurable. A thermally activated process for the diffusion of atomic hydrogen on Si(111)7x7 was found, with a barrier of 1.5±0.2 eV, and a pre-exponential factor of 10^{-3} cm^{2}s^{-1} [65]. In a recent study, phonon-assisted sticking of molecular hydrogen on Si(111)7x7 has been identified, with an effective barrier to dissociative adsorption of 0.9±0.1 eV [67].

Isothermal desorption of hydrogen from singular Si(001)2x1 at low coverages has also been followed by SHG [68]. The enhanced sensitivity of the SH response from Si(001)2x1 to adsorption, when using 1064 nm excitation (Sect. 8.3), was again exploited. An effective π-bond strength of 0.25 eV, a desorption activation energy of 2.5±0.1 eV, and a first-order pre-exponential factor of 10^{15} s^{-1}, was deduced [68]. The kinetics of oxygen dissociation on Si(111)7x7 has also been followed, which has produced evidence of more than one dissociation channel and/or molecular oxygen species being present on the surface [69]. In another study, evidence of surface defects created by redesorption from an atomic oxygen state, was found [70].

8.4.4 The Si/SiO$_2$ Interface

The technological importance of this interface has encouraged intensive study by SHG. Shen and co-workers, in an early study, showed that surface and bulk contributions from native-oxide-covered Si(111) and Si(001) wafers could be identified [20]. It is clear from the previous section that the formation of the oxide interface will remove the surface state resonances responsible for the enhanced signal when 1064 nm excitation is used and, indeed, no resonantly-enhanced signal is observed from the Si/SiO_2 interface, using either 1064 nm or 532 nm excitation. The surface and bulk, higher-order, contributions are comparable in size and detailed studies are required to separate them. Indeed, for some terms, exact separation is not possible [18].

Vicinal $Si(111)/SiO_2$ interfaces, where the χ^s_{xxx} component is large, have been used to help quantify surface and bulk contributions [71, 72]. Recently, detailed studies by van Driel and co-workers of vicinal Si/SiO_2 interfaces, with oxide produced by both dry and wet oxidation, and using 765 nm excitation from a mode-locked Ti:sapphire laser, have been reported [55, 73]. These are the highest quality data yet obtained from the Si/SiO_2 interface, and show the potential of the new pulsed laser sources now available. Similar studies, with comparable data quality, have been performed, using 1053 nm excitation, on chemically-modified vicinal Si(111) interfaces, which reveal a correlation between SH response and the density of interface traps [74, 75]. The azimuthal rotation plots (Sect. 8.1) of van Driel and co-workers revealed the presence of C_{1v} and C_{3v} symmetry components at the interface, associated with the step structure, and also a C_{2v} symmetry component after the dry oxidisation, which is consistent with the presence of dimers at the interface for low temperature treatment, and with the tridymite form of c-SiO_2 at high temperatures [55, 73].

Of particular interest in the second publication [75] is the similar size of the surface and bulk contributions, and the mainly real tensor components, both of which indicate off-resonance behaviour. Recently, Daum and co-workers have identified a strong resonance at 3.3 eV, for both clean and oxidised Si surfaces [38, 76]. In their spectra of singular Si(001)2x1 (Fig. 57), the SH intensity at 3.24 eV (the SH energy for 765 nm excitation) is near the resonance maximum. This apparent conflict was resolved by McGilp and co-workers in a higher resolution study which showed that, for p-in/p-out polarisation, the resonance peak of Si(001)2x1 is a doublet (Figs. 8.8, 8.9), with the 3.24 eV SH energy of van Driel and co-workers lying between the resonance peaks [33]. This will be discussed further in the next section.

The SH response from the Si/SiO_2 system is also sensitive to static electric fields at the interface (electric-field-induced SHG [77]), and inhomogeneous strain [78], as has been demonstrated by Aktsipetrov, Kulyuk and co-workers [79-81]. Information on the technologically important metal-oxide-semiconductor (MOS) structure has been obtained recently [81]. The $Si(111)/SiO_2$/electrolyte interface [82, 83], and the $Si(110)/SiO_2$ interface [84], have also been studied.

8.5 Spectroscopic SHG and SFG

The spectroscopic potential of SHG is clear when the quite different responses of the Si(111)-Ga [29], Si(111)-Sb [30] and Si(111)-As [32] systems at two different excitation energies are compared. Reliable, pulsed OPO and Ti:sapphire systems, tuneable over a wide range, are now becoming available and a substantial increase in the number of spectroscopic three-wave mixing experiments may be expected. The first *spectroscopic* SHG and SFG study of a semiconductor interface was of the buried $CaF_2/Si(111)$ interface through 50 nm of CaF_2 [35]. Resonant three-wave mixing was used to determine an interface state band gap. Fig. 8.6 shows that the dispersion of the SF signal matches the SH signal, when plotted as a function of dye laser frequency. The resonance must thus occur at the fundamental frequency of the dye laser, rather than at the SH frequency. This allows the bandgap at the buried interface to be determined [35]. This is an excellent example of how an epioptic probe can provide unique information about the structure and properties of a buried interface between an insulator and a semiconductor. The main spectral feature at 2.4 eV in Fig. 8.6 was assigned to direct transitions at the two-dimensional band gap, while the weaker feature at 2.25 eV was tentatively assigned to a two-dimensional exciton.

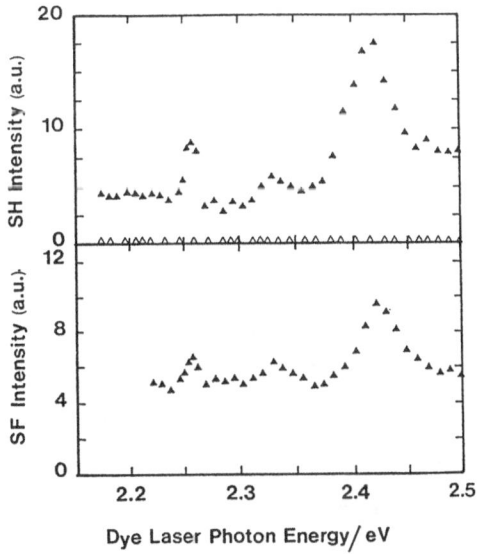

Fig. 8.6. Resonant three-wave mixing, as a function of dye laser photon energy, from epitaxial $CaF_2/Si(111)$ (after [35]).

This type of experiment has remained difficult due to a lack, until recently, of broad band pulsed laser sources. Another material system with resonances accessible by dye laser was studied by Yodh and co-workers. In an elegant series of experiments, the buried ZnSe/GaAs(001) interface was probed and, by combining spectroscopic SHG, SFG and photoinduced band bending, a resonance associated with a quantum well (QW) state at the buried interface was identified [36, 85]. Sensitivity to interfacial electronic traps, lattice relaxation and buried interface reconstruction have also been demonstrated in this interesting system [37]. The effect of the depletion layer electric field (band bending) on SHG from GaAs(001) has also been studied by Yodh and co-workers [86].

As mentioned above, Daum and co-workers, using an OPO-based system, have found a resonance just below the direct optical gap of Si [38, 76]. Figure 8.7 shows the resonant SFG and SHG behaviour of oxidised, clean and hydrogen-terminated singular Si(001) samples. Similar results were found for Si(111)

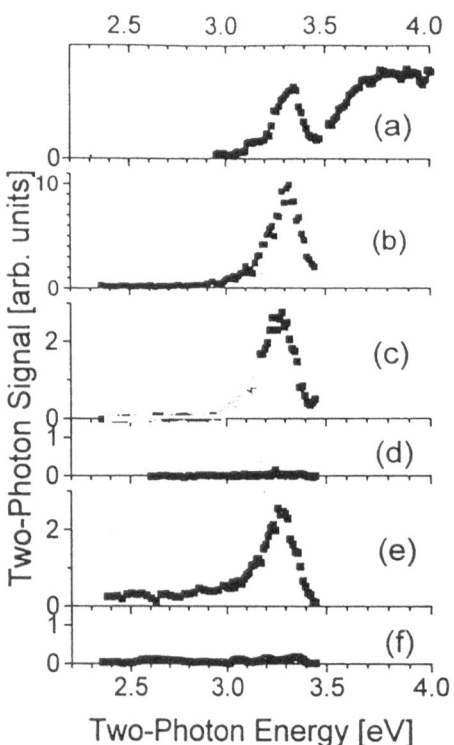

Fig. 8.7. SFG (a) and SHG (b)-(f) spectra of differently oxidised (a)-(d), clean Si(001) 2x1 (e), and hydrogen-terminated (f) Si(001) samples [38].

samples. The bulk response is negligible in comparison with the resonance response, which is dominated by the χ^{s}_{zzz} component [38]. The resonance was ascribed to direct transitions between valence and conduction band states in a few monolayers of strained Si at the surface or interface. The hydrogen-terminated sample is neither reconstructed nor strained, and shows no resonance [38].

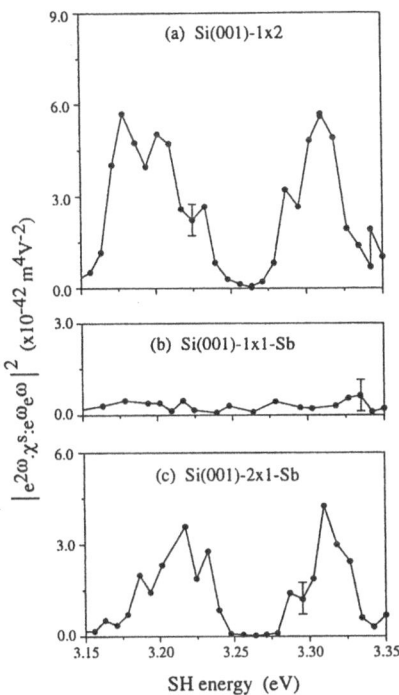

Fig. 8.8. Variation of p-in/p-out SH intensity with SH energy, for *singular* (a) Si(001)1x2, (b) Si(001)1x1-Sb, and (c) Si(001)2x1-Sb [33].

McGilp and co-workers have studied this resonance on singular and vicinal Si(001) and Si(001)-Sb [33, 87]. The resonance is revealed to be a doublet for p-in/p-out polarisation (Fig. 8.8a), which explains the contrasting results of van Driel and co-workers [55], as discussed above. The Si(001)1x1-Sb result (Fig. 8.8b), showing no significant resonance behaviour in contrast to the Si(001)2x1-Sb system, is also of interest. Without detailed intensity/voltage studies, LEED cannot readily distinguish between a (1x1) adsorbate monolayer on a (1x1) substrate, and an adsorbate layer *disordered* in the surface plane on such a substrate. In each case a (1x1) LEED pattern will be observed. The Si(001)-Sb results show that SHG can distinguish between these cases, as long as there is some isotropic component in the response. In Fig. 8.8, the SH response is dominated by χ^{s}_{zzz}, which is isotropic in the surface plane. The Si(001)1x1-Sb structure cannot comprise disordered (2x1) domains of Sb dimers, because the (1x1)-Sb response would then be similar to the (2x1)-Sb response (Fig. 8.8c).

These results support the conclusions of Cricenti and co-workers [88] that the (1x1)-Sb structure is unusual, having isolated Sb atoms on neighbouring Si atoms, the Sb atoms only forming dimers on further annealing.

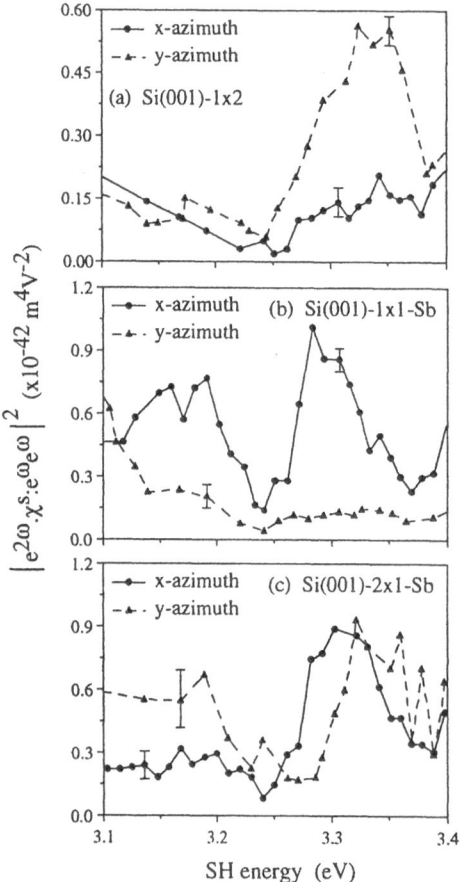

Fig. 8.9. Variation of q-in/s-out SH intensity with SH energy, for *vicinal* (a) Si(001)1x2, (b) Si(001)1x1-Sb, and (c) Si(001)2x1-Sb [33]. Note the large in-plane anisotropy in (a) and (b), and the switch of intensity between azimuths.

The vicinal Si(001)-Sb system is also of interest. Figure 8.9 shows a dramatic switch of SH resonance intensity from the y-azimuth ($[\bar{1}10]$ direction) of Si(001)1x2 to the x-azimuth ($[110]$ direction) of Si(001)1x1-Sb. Symmetry analysis shows that the 1m mirror plane perpendicular to the step edges, present in all previous studies of vicinal systems, has been lost. A reconstruction at the step edge is occurring on Sb adsorption which, on annealing to higher temperatures, forces the formation of the opposite domain, Si(001)2x1-Sb terrace structure as shown by LEED, in contrast to the original, Si(001)1x2 terrace structure. Recently, Petukhov and Liebsch have shown that the origin of the large SH anisotropy of vicinal Al(001) is consistent with the nonsymmetric stacking of bulk lattice planes parallel to the macroscopic surface, rather than the nonsymmetric electron density corrugation at the stepped surface [89]. In contrast, the absence of the 1m mirror plane shows that, for Si(001)1x1-Sb, the resonance response is dominated by the surface steps [33]. Not only is SHG sensitive to surface *terrace* reconstruction, as shown by Heinz and co-workers [4] (discussed in Sect. 8.3), but *spectroscopic* SHG can be sensitive to the local reconstruction of surface *steps*, if electronic resonances localised at the step edges can be probed [33].

8.6 SFG for Vibrational Spectroscopy of Surfaces

The use of SFG to determine whether an electronic resonance is associated with the fundamental or SH frequency has been mentioned above. However, the most important development has been IR-visible SFG, which allows vibrational spectroscopy of surfaces and interfaces to be performed by resonant IR excitation [90-93]. In this type of experiment, tuneable IR radiation is mixed with visible radiation to obtain a visible SF signal which shows resonant enhancement at vibrational mode frequencies.

Hydrogen on Si surfaces has been studied in some detail. Picosecond SFG of Si(111)1x1-H, as a function of temperature, revealed a lifetime of 0.8±0.1 ns for the Si-H stretch vibration [94]. SFG is a coherent process and this allows dephasing of the transient polarisation of this vibration to be followed [95]. A two-phonon bound state for this vibration has also been identified [96] and the effect of carrier density in the substrate investigated [97]. Hydrogen-terminated vicinal Si(111) surfaces have also been studied [98-101]. Measurements of the excited-state lifetimes of the Si-H stretch vibration show that the steps play an important role in vibrational energy transfer at the surface. Studies of monohydride- and dihydride-terminated steps show that inter-adsorbate energy transfer competes efficiently with energy relaxation to the substrate, for this system [99, 100].

These elegant studies show the detailed information available from SFG of vibrational modes at surfaces and, with the more general availability of appropriate picosecond sources, much more work in this area is to be expected.

8.7 Semiconductor Growth

In contrast to the exciting results obtained using RAS and SE, there have only been a few studies of semiconductor growth using nonlinear optical techniques.

8.7.1 Elemental Semiconductors

In an early study, Heinz and co-workers, using 1064 nm excitation, showed that Si deposition on Si(111)7x7 at room temperature produced a decrease in the SH intensity, while the signal was constant under epitaxial growth conditions at 650°C [102]. Under these experimental conditions, the SH signal is sensitive to disorder in the surface layers. Similar behaviour was observed by Hollering and co-workers for room temperature growth of Si on Si(001) in a simultaneous RHEED and SHG study [103]. However, Ge was observed to produce an initial decrease, followed by an increase up to 1ML, before decreasing again. This was interpreted as the effect of the formation of a Si-Ge dipolar layer at the interface. Epitaxial growth of Si and Ge at high temperatures gave a constant SH signal for Si, but Ge again showed a maximum around 1 ML [103].

8.7.2 Compound Semiconductors

Semiconductors of zinc blende structure are *acentric*, and thus have a *bulk* dipole response. However, Shen and co-workers showed that, by appropriate choice of polarisations and substrate orientation, the bulk dipole response can be suppressed, while allowing many of the surface tensor components to be determined [104]. Surface SHG was observed from Sn adsorbed on GaAs(001), showing that surface SHG studies are not restricted to centrosymmetric systems [104]. The interesting results of Yodh and co-workers from the ZnSe/GaAs(001) interface were discussed in Sect. 8.5 [36, 37, 85, 86].

As regards growth, Pemble and co-workers observed a large SH intensity variation during the adsorption of TMG on GaAs(001) [105], while in an MBE growth study, Yamada and Kimura observed rotational anisotropy which differed between the GaAs(001)2x1 and the GaAs(001)4x6 surface reconstructions [106]. It remains to be seen whether nonlinear optical techniques can offer any advantages over the simpler linear optical techniques in the semiconductor growth area.

References

1. Shen, Y.R.: The Principles of Non-linear Optics. Wiley, New York 1984
2. Shen, Y.R.: Journal of Vacuum Science and Technology B 3, 1464 (1985)
3. Tom, H.W.K., Mate, C.M., Zhu, X.D., Crowell, J.E., Heinz, T.F., Somorjai, G. A., Shen, Y.R.: Physical Review Letters 52, 348 (1984)
4. Heinz, T.F., Loy, M.M.T., Thompson, W.A.: Physical Review Letters 54, 63 (1985)
5. McGilp, J.F., Yeh, Y.: Solid State Communications 59, 91 (1986)
6. McGilp, J.F.: Semiconductor Science and Technology 2, 102 (1987)
7. McGilp, J.F.: Journal of Physics: Condensed Matter 1, SB85 (1989)
8. Richmond, G.L., Robinson, J.M., Shannon, V.L.: Progress in Surface Science 28, 1 (1988)
9. Shen, Y.R.: Annual Reviews of Physics and Chemistry 40, 327 (1989)
10. Heinz, T.F., Reider, G.A.: Trends in Analytical Chemistry 8, 235 (1989)
11. Heinz, T.F. in: Ponath H.-E., Stegeman, G.I. (eds.) Nonlinear Surface Electromagnetic Phenomena. North-Holland, Amsterdam 1991, p. 353
12. Guyot-Sionnest, P., Chen, W., Shen, Y.R.: Physical Review B 33, 8254 (1986)
13. Sipe, J.E., Moss, D.J., van Driel, H.M.: Physical Review B 35, 1129 (1987)
14. Hopf, F.A., Stegeman, G.I.: Applied classical electrodynamics, vols 1 and 2. Wiley, New York 1986
15. Bloembergen, N., Chang, R.K., Jha, S.S., Lee, C.H.: Physical Review 174, 813 (1968)
16. Heinz, T.F.: PhD dissertation, University of California, Berkeley, California 1982
17. Mizrahi, V., Sipe, J.E.: Journal of the Optical Society of America B 5, 660 (1988)
18. Sipe, J.E., Mizrahi, V., Stegeman, G.I.: Physical Review B 35, 9091 (1987)
19. Guidotti, D., Driscoll, T.A., Gerritsen, H.J.: Solid State Communications 46, 337 (1983)
20. Tom, H.W.K., Heinz T.F., Shen, Y.R.: Physical Review Letters 51, 1983 (1983)
21. Litwin, J.A., Sipe J.E., van Driel, H.M.: Physical Review B 31, 5543 (1985)
22. Giordmaine, J.A.: Physical Review 138, A 1599 (1965)
23. Tom, H.W.K., Aumiller, G.D.: Physical Review B 33, 8818 (1986)
24. Tom, H.W.K.: PhD dissertation, University of California, Berkeley, California 1984
25. Aktsipetrov, O.A., Baranova, I.M., Il'inskii, Y.A.: Soviet Physics-JETP 64, 167 (1986)
26. Tang, Z.-R., Cavanagh, M., McGilp, J.F.: Journal of Physics: Condensed Matter 5, 3791 (1993). Ibidem 5, 7903 (1993)
27. Hollering, R.W.J., Dijkkamp, D., Lindelauf, H.W.L., van der Heide, P.A.M., Krijn, M.P.C.M.: Journal of Vacuum Science and Technology A 8, 3997 (1990)
28. O'Mahony, J.D., Kelly P.V., McGilp, J.F.: Applied Surface Science 56-8, 449 (1992)
29. Kelly, P.V., O'Mahony, J.D., McGilp J.F., Rasing, Th.: Surface Science 269/270, 849 (1992)
30. Power J.R., McGilp, J.F.: Surface Science 287/288, 708 (1993)
31. O'Mahony, J.D., McGilp, J.F., Verbruggen, M.H.W., Flipse, C.F.J.: Surface Science 287/288, 713 (1993)
32. Power, J.R., McGilp, J.F.: Surface Science 307/309, 1066 1994)
33. Power, J.R., O'Mahony, J.D., Chandola, S., McGilp, J.F.: Physical Review Letters, in press (1995)
34. Chang, R.K., Ducuing, J., Bloembergen, N.: Physical Review Letters 15, 6 (1965)
35. Heinz, T.F., Himpsel, F.J., Palange, E., Burstein, E.: Physical Review Letters 63, 644 (1989)
36. Yeganeh, M.S., Qi, J., Yodh, A.G., Tamargo, M.C.: Physical Review Letters 68, 3761 (1992)
37. Yeganeh, M.S., Qi, J., Yodh, A.G., Tamargo, M.C.: Physical Review Letters 69, 3579 (1992)

204   J. F. McGilp

38. Daum, W., Krause, H.-J., Reichel, U., Ibach, H.: Physical Review Letters *71*, 1234 (1993)
39. Boyd, G.T., Shen, Y.R., Hansch,T.W.: Optics Letters *11*, 97 (1986)
40. Krause, H.-J., Daum, W.: Applied Physics Letters *60*, 2180 (1992)
41. Krause, H.-J., Daum, W.: Applied Physics *B 56*, 8 (1993)
42. Kelly, P.V., Tang, Z.-R., Woolf, D.A., Williams, R.H., McGilp, J.F.: Surface Science *251/2*, 87 (1991)
43. Maker, P.D., Terhune, R.W., Nisenoff, M., Savage, C.M.: Physical Review Letters *8*, 21 (1962)
44. Weber, M. J.: Handbook of Laser Science and Technology, vol.III, part 1. CRC Press, Boca Raton 1986, p. 44
45. Heinz, T.F., Loy, M.M.T., Thompson, W.A.: Journal of Vacuum Science and Technology *B 3*, 1467 (1985)
46. Takayanagi, K., Tanishiro, Y., Takahashi, M., Takahashi, S.: Journal of Vacuum Science and Technology *A 3*, 1502 (1985)
47. Elwijk, H.B., Dijkkamp, D., van Loenen, E.J.: Physical Review *B 44*, 3802 (1991)
48. Noguez, C., Shkrebtii, A.I., Del Sole, R.: Surface Science, in press (1995)
49. Pedersen, K.: to be published
50. Suzuki, T., Hirabayashi, Y.: Japanese Journal of Applied Physics *32*, L610 (1993)
51. Höfer, U., Li, L., Ratzlaff, E.H., Heinz, T.F.: Optical Society of America Technical Digest Series *12*, 141 (1993)
52. McGilp J.F., Power, J.R.: to be published
53. Jiang, H.B., Liu, Y.H., Liu, X.Z., Wang, W.C., Zheng, J.B., Zhang, Z.M.: Applied Physics Letters *65*, 1558 (1994)
54. Hollering, R.W.J., Dijkkamp, D., Elswijk, H.B.: Surface Science *243*, 121 (1991)
55. Lüpke, G., Bottomley, D.J., van Driel, H.M.: Journal of the Optical Society of America *B 11*, 33 (1994)
56. Pedersen, K., Morgen, P.: Phyisca Scripta *T54*, 238 (1994)
57. Chen, J.M., Bower, J.R., Wang, C.S., Lee, C.H.: Optics Communications *9*, 132 (1973)
58. Kelly, P.V., O'Mahony, J.D., McGilp J.F., Rasing, Th.: Applied Surface Science *56-58*, 453 (1992)
59. Nogami, J., Park, S., Quate, C.F.: Surface Science *203*, L631 (1988)
60. Kawazu, A., Sakama, H.: Physical Review *B 37*, 2704 (1988)
61. Patel, J.R., Zegenhagen, J., Freeland, P.E., Hybertsen, M.S., Golovchenko, J.A., Chen, D.M.: Journal of Vacuum Science and Technology *B 7*, 894 (1989)
62. Lander, J.J., Morrison, J.: Surface Science *2*, 553 (1964)
63. Nicholls, J.M., Reihl, B., Northrup, J.E.: Physical Review *B 35*, 4137 (1987)
64. Reider, G.A., Höfer, U., Heinz, T.F.: Journal of Chemical Physics *94*, 4080 (1991)
65. Reider, G.A., Höfer, U., Heinz, T.F.: Physical Review Letters *66*, 1994 (1991)
66. Zhu, X.D., Rasing, Th., Shen, Y.R.: Physical Review Letters *61*, 2883 (1988)
67. Bratu, P., Höfer, U.: to be published
68. Höfer, U., Li, L., Heinz, T.F.: Physical Review *B 45*, 9485 (1992)
69. Bratu, P., Kompa, K.L., Höfer, U.: Physical Review *B 49*, 14070 (1994)
70. Pedersen, K., Morgen, P.: to be published
71. van Hasselt, C.W., Verheijen, M.A., Rasing, Th.: Physical Review *B 42*, 9263 (1990)
72. Verheijen, M.A., van Hasselt, C.W., Rasing, Th.: Surface Science *251*, 467 (1991)
73. Lüpke, G., Bottomley, D.J., van Driel, H.M.: Physical Review *B 47*, 10389 (1993)
74. Bjorkman, C.H., Yasuda, T., Shearon, Jr., C.E., Ma, Y., Lucovsky, G. Emmerichs, U., Meyer, C., Leo, K., Kurz, H.: Journal of Vacuum Science and Technology *B 11*, 1521 (1993)

75. Emmerichs, U., Meyer, C., Bakker, H.J., Kurz, H., Bjorkman, C.H., Shearon, Jr., C.E., Ma, Y., Yasuda, T., Jing, Z., Lucovsky, G., Whitten, J.L. Physical Review *B 50*, 5506 (1993)
76. Daum, W., Krause, H.-J., Reichel, U., Ibach, H.: Phyisca Scripta *T49 B*, 513 (1993)
77. Lee, C.H., Chang, R.R., Bloembergen, N.: Physical Review Letters *18*, 167 (1967)
78. Govorkov, S.V., Emel'yanov, V.I., Koroteev, N.I., Petrov, G.I., Shumay, I.L., Yakovlev, V.V.: Journal of the Optical Society of America *B 6*, 1117 (1989)
79. Kulyuk, L.L., Shutov, D.A., Strumban, E.E., Aktsipetrov, O.A.: Journal of the Optical Society of America *B 8*, 1766 (1991)
80. Kravetsky, I.V., Kulyuk, L.L., Micu, A.V., Shutov, D.A., Strumban, E.E., Cobianu, C., Dascalu, D.: Applied Surface Science *63*, 269 (1993)
81. Aktsipetrov, O.A., Fedyanin, A.A., Golovkina, V.N., Murzina, T.V.: Optics Letters *19*, 1 (1994)
82. Fischer, P.R., Daschbach, J.L., Richmond, G.L.: Chemical Physics Letters *218*, 200 (1994)
83. Fischer, P.R., Daschbach, J.L., Gragson, D.E., Richmond, G.L.: Journal of Vacuum Science and Technology *A 12*, 2617 (1994)
84. Malliaras, G.G., Wierenga G.G., Rasing, Th.: Surface Science *287/288,* 703 (1993)
85. Yeganeh, M.S., Qi, J., Culver, J.P., Yodh, A.G., Tamargo, M.C.: Physical Review *B 49*, 11196 (1994)
86. Qi, J., Yeganeh, M.S., Koltover, I., Yodh, A.G., Theis, W.M.: Physical Review Letters *71*, 633 (1992)
87. McGilp, J.F., Cavanagh, M., Power, J.R., O'Mahony, J.D.: Applied Physics *A 59*, 401 (1994)
88. Cricenti, A., Selci, S., Felici, A.C., Ferrari, L., Contini, G., Chiarotti, G.: Physical Review *B 47*, 15745 (1993)
89. Petukhov, A.V., Liebsch, A.: Surface Science, in press (1995)
90. Guyot-Sionnest, P., Hunt, J.H., Shen, Y.R.: Physical Review Letters *59*, 1597 (1987)
91. Zhu, X.D., Suhr H., Shen, Y.R.: Physical Review *B 35*, 3047 (1987)
92. Harris, A.L., Chidsey, C.E.D., Levinos, N.J., Loiacono, D.N.: Chemical Physics Letters *141*, 350 (1987)
93. Guyot-Sionnest, P., Superfine, R., Hunt, J.H., Shen, Y.R.: Chemical Physics Letters *144*, 1 (1988)
94. Chabal, Y.J., Dumas, P., Guyot-Sionnest,P., Higashi, G.S.: Surface Science *242*, 524 (1991)
95. Guyot-Sionnest, P.: Physical Review Letters *66*, 1489 (1991)
96. Guyot-Sionnest, P.: Physical Review Letters *67*, 2323 (1991)
97. Guyot-Sionnest, P.: Journal of Electron Spectroscopy and Related Phenomena *64/65*, 1 (1993)
98. Morin, M., Jakob, P., Levinos, N.J., Chabal Y.J., Harris, A.L.: Journal of Chemical Physics *96,* 6203 (1992)
99. Kuhnke, K., Morin, M., Jakob, P., Levinos, N.J., Chabal, Y.J., Harris, A.L.: Journal of Chemical Physics *99*, 6114 (1993)
100. Morin, M., Kuhnke, K., Jakob, P., Chabal, Y.J., Levinos, N.J., Harris, A.L.: Journal of Electron Spectroscopy and Related Phenomena *64/65*, 11 (1993)
101. Kuhnke, K., Harris, A.L., Chabal, Y.J., Jakob, P., Morin, M.: Journal of Chemical Physics *100*, 6896 (1994)
102. Heinz, T.F., Loy, M.M.T., Iyer, S.S.: Materials Research Society Symposium Proceedings *55*, 697 (1987)
103. Hollering, R.W.J., Hoeven, A.J., Lenssinck, J.M.: Journal of Vacuum Science and Technology *A 8*, 3194 (1990)
104. Stehlin, T., Feller, M., Guyot-Sionnest, P., Shen, Y.R.: Optics Letters *13*, 389 (1988)

105. Buhaenko, D.S., Francis, S.M., Goulding, P.A., Pemble, M.E.: Journal of Crystal Growth *97*, 595 (1989)
106. Yamada, C., Kimura, T.: Physical Review Letters *70*, 2344 (1993)

Chapter 9. Conclusions

John McGilp

Department of Physics, Trinity College, Dublin 2, Ireland

Epioptics is now well established, and is making a major contribution to the study of surfaces and interfaces in general, and those of semiconductors in particular. Significant advantages over conventional surface spectroscopies have been demonstrated: all pressure ranges and transparent media are accessible; insulators can be studied without the problem of charging effects; buried interfaces can be studied due to the large penetration depth of optical radiation. Epioptic techniques offer micron lateral resolution and femtosecond temporal resolution. Non-destructive, *in situ* characterisation of thin films, surface and interfaces in all pressure régimes is central to the development of new materials and processes, particularly in this evolving era of nanoscale structures. It is these advantages over conventional surface spectroscopies which have driven the development of epioptics.

In the next few years, a major effort is expected in the theoretical area, as there is now a wealth of high quality experimental data available. A detailed understanding of the linear optical response may come quite rapidly, but calculating the nonlinear response of anything but the simplest surface and interface systems remains a daunting computational problem.

There are three main approaches, at present, to evaluating the linear optical response: these can be categorised as band structure, bond model and equation-of-motion methods. Of the various band structure approaches, self-consistent pseudopotential calculations are very promising, and appear to offer superior results to the tight-binding method. The bond approach, which uses arrays of polarisable atoms, appears to be useful in calculating the vibrational response, and there is interest in attempting to extend this approach to evaluating the electronic response at optical frequencies. The equation-of-motion method may yet turn out to be the best way forward for complex surface and interface systems.

On the experimental side, the linear optical techniques (SE, RAS, SDR, SPA) are now well-developed, and are beginning to be widely used for semiconductor interface characterisation and growth monitoring. Using SE, the dielectric function of a thin layer on a semiconductor, together with the layer thickness, can be determined from the measured effective dielectric function, using the three-phase model, for thicknesses down to a few monolayers. The dielectric function contains information about the electronic structure of the surface or the layer, and is changed by surface reconstruction, strain, temperature and roughness. In the monolayer region and below, only the product of thickness and dielectric function obtained from the three-phase model continues to be a useful quantity. This is a limit of the model, and not a fundamental limit to the sensitivity of the technique. Changes in the effective dielectric function due to coverages below a monolayer can easily be detected using photometric ellipsometers.

The other linear optical techniques (RAS, SDR, SPA) are, experimentally, somewhat simpler and more direct, in that they measure changes in reflectance due to variation in the surface or layer dielectric function with azimuthal angle (RAS), or on adsorption (SDR, SPA). However, these techniques usually provide less information than a full SE study. Their advantage lies mainly in the area of adsorption and growth monitoring, where it is often sufficient to identify a signature characteristic of the desired process.

For all these linear optical techniques, a model interface would be highly desirable for instrumental calibration purposes and standardisation between different laboratories and materials growth facilities. Strain effects in optical components and windows are a particular problem. The hydrogenated vicinal Si(001) surface is a possible candidate for an interface standard [1]. There may also be a sign convention problem to be resolved for these linear optical techniques [2]. As regards instrumentation, various double modulation spectroscopies are being developed for the linear optical area, where a modulation at low frequency is added to the high frequency photoelastic modulation of the polarisation. An example is photoellipsometry, which can be used to measure surface photovoltage [3].

Raman spectroscopy (RS) can show submonolayer sensitivity, provided the adsorbed layer possesses Raman-active phonon modes at different frequencies to those of the substrate, thus allowing discrimination against the large substrate signals. Instrumentation of the required sensitivity is now commercially available. Selection of the laser excitation frequency is important, as resonant enhancement of the Raman cross-section in the region of a surface or interface optical transition can aid both sensitivity and the ability to discriminate against the substrate signal. The development of broad band CW laser sources is needed in this area. A major advantage of RS is that it provides a fingerprint of interface composition and chemistry via characteristic vibrational frequencies. Of the other techniques discussed, only SFG, which requires more complex pulsed laser systems, offers this possibility. Such direct information on interface structure and reactivity is particularly suited to heterostructure growth monitoring and control.

The major application of PL as an epioptic technique concerns specially grown semiconductor structures and, in particular, low dimensional systems. Two-dimensional structures have been emphasised, but quantum wire and quantum dot structures also offer interesting opportunities for PL.

The nonlinear spectroscopies, SHG and SFG, provide complementary information to those of the linear optical techniques. SHG and SFG give very small signals, but these can be dominated by the surface and interface response, while the linear optical techniques give large signals, but only a small fraction of the signal contains surface and interface information. The recent commercial availability of a new generation of broad band pulsed laser sources may be expected to accelerate the development of nonlinear spectroscopy at surfaces and interfaces. Quantitative measurements are to be encouraged in the SHG area, as these allow comparison with theoretical work. The use of phenomenological models to extract experimental $\chi_{ijk}^{(2)}$ values, however, is more questionable than in the linear case. Full evaluation of the behaviour of the EM fields in the surface or interface region needs to be included in the complete calculation of the nonlinear optical response. This is one of the interesting challenges to be faced in the future development of epioptics.

References

1. Müller, A.B., Reinhardt, F., Richter, W., Rose, K.C., Rossow, U.: Thin Solid Films *233*, 19 (1993)
2. Wijers, C.M.J.: to be published
3. Drévillon, B.: to be published

Appendix A. SHG Formalism

Michele Cini

Dipartimento di Fisica, II Università di Roma, 'Tor Vergata', Via della Ricerca Scientifica, I-00133 Roma, Italia

A.1 Second Harmonic Generation: Perturbation Theory

The Green's function formulation that we propose for the process of sum frequency generation (SFG), $\omega_3 = \omega_1 + \omega_2$, is tailored for interface problems, and includes the dielectric response of the system and the A^2 term. Let V_A be the photon absorption operator. In the place of the photon emission operator we use

$$V_E(\omega_3) = (1/c)\sum_k \int_0^\infty dz' D_{ik}^R(\omega_3, Q, z, z') J_k(Q, z') \tag{A.1}$$

Here, z is the distance of the detector from the interface, D_{ik}^R is the retarded Green's function [1-3] accounting for polarisation and absorption effects, and J is the current operator. Then, the ith component of the vector potential of the SFG field, in the independent-electron approximation, is given by:

$$
A_{if} = -\frac{2}{\hbar^3} X(\omega_1, \omega_1) + \frac{e^2}{2\hbar^2 mc^2} \sum_{m,n} f(m)[1 - f(n)]\left\{ \frac{[V_E(\omega_3)]_{mn}[A^2(\omega_1)]_{nm}}{\omega_{nm} - \omega_1 - i0} \right.
$$
$$
+ \frac{[A^2(\omega_1)]_{mn}[V_E(\omega_3)]_{nm}}{\omega_{nm} + \omega_3 - i0} + 2\frac{[A(\omega_1)\cdot A(\omega_3)]_{nm}[V_A(\omega_1)]_{mn}}{\omega_{nm} + \omega_1 - i0}
$$
$$
\left. + 2\frac{[A(\omega_1)\cdot A(\omega_3)]_{mn}[V_A(\omega_1)]_{nm}}{\omega_{nm} - \omega_1 - i0} \right\}
$$

$$\tag{A.2}$$

where the summations are due to the diamagnetic term, and require a self consistent treatment, and

$$X(\omega_1, \omega_2) =$$

$$\sum_s f(s) \sum_n [1 - f(n)] \sum_r [1 - f(r)] \left\{ \frac{[V_E(\omega_3)]_{sn}[V_A(\omega_2)]_{nr}[V_A(\omega_1)]_{rs}}{(\omega_{rs} - \omega_1 - i0)(\omega_{ns} - \omega_1 - \omega_2 - i0)} \right.$$

$$+ \frac{[V_A(\omega_2)]_{sn}[V_E(\omega_3)]_{nr}[V_A(\omega_1)]_{rs}}{(\omega_{rs} - \omega_1 - i0)(\omega_3 - \omega_1 + \omega_{ns} - i0)} + \frac{[V_A(\omega_2)]_{sn}[V_A(\omega_1)]_{nr}[V_E(\omega_3)]_{rs}}{(\omega_3 + \omega_{rs} - i0)(\omega_3 - \omega_1 + \omega_{ns} - i0)}$$

$$- \sum_s f(s) \sum_n [1 - f(n)] \sum_m f(m) \left\{ \frac{[V_E(\omega_3)]_{mn}[V_A(\omega_2)]_{sm}[V_A(\omega_1)]_{ns}}{(\omega_{ns} - \omega_1 - i0)(\omega_{nm} - \omega_1 - \omega_2 - i0)} \right.$$

$$+ \frac{[V_A(\omega_2)]_{mn}[V_E(\omega_3)]_{sm}[V_A(\omega_1)]_{ns}}{(\omega_{ns} - \omega_1 - i0)(\omega_3 - \omega_1 - \omega_{mn} - i0)} + \left. \frac{[V_A(\omega_2)]_{mn}[V_A(\omega_1)]_{sm}[V_E(\omega_3)]_{ns}}{(\omega_3 + \omega_{ns} - i0)(\omega_3 - \omega_1 - \omega_{nm} - i0)} \right\}$$

$$\tag{A.3}$$

The Fermi distribution at the one-electron eigenvalue, ε_n, is $f(n)$, and $\omega_{rs} = \varepsilon_r - \varepsilon_s$.

A.2 The Many-Photon Approach

Let us assume that we have M modes of given frequency, ω_0, and polarisation, ε, in the exciting beam. For each mode, there is a matter-radiation coupling term in the Hamiltonian,

$$H'_k = \sum_{m \neq n} \sum_\sigma M^{(k)}_{mn} a^\dagger_{m\sigma} a_{n\sigma} (b_k + b^\dagger_k) \tag{A.4}$$

where in the dipole approximation

$$M^{(k)}_{mn} = (e/mc) <m|A^{(k)} \cdot P|n> = (e/mc)A^{(k)} <m|\varepsilon \cdot P|n> \tag{A.5}$$

Hence, H'_k factorizes in the form

$$H'_k = A^{(k)} L(b_k + b^\dagger_k) \tag{A.6}$$

where L operates on electron but not on photon degrees of freedom. The part of the Hamiltonian which describes the field and its interaction with the matter then reads

$$H = \sum_{k}^{M} \omega_0 b_k^\dagger b_k + A^{(k)} L(b_k + b_k^\dagger) \tag{A.7}$$

It can be shown that we can always replace the M modes, with couplings $A^{(k)}$, by one effective mode with coupling $\|A\| = (\sum_k [A^{(k)}]^2)^{1/2}$ if the (arbitrary) normalization volume, V, is taken to be the laser cavity volume.

Accordingly, consider an electronic system interacting with a one-mode laser radiation field. In general, we may consider several interacting electrons described by the Hamiltonian, H_e. Let N be the dimensionality of the Hilbert space, and $\{|i\rangle, i = 1, .. N\}$ a basis set. The free radiation field has two modes of frequencies ω_0 and v, and is described by the Hamiltonian

$$H_{ph} = \omega_0 b^\dagger b + v b'^\dagger b' \tag{A.8}$$

Here b refers to photons of frequency ω_0 from the exciting laser beam and b' to the scattered photons. The electron-photon interaction terms are

$$H_I = M(b + b^\dagger) \tag{A.9}$$

$$H_I' = M'(b' + b'^\dagger) \tag{A.10}$$

where M and M' are operators acting only on the electron degrees of freedom. Thus, the total Hamiltonian is

$$H = H_e + H_{ph} + H_I + H_I' \tag{A.11}$$

and can be thought of as implying a summation over all possible scattered photon modes. The exciting photons belong to one mode and are in a coherent state, $|c\rangle$, such that $d|c\rangle = 0$ with $d = b - \beta$. In the d representation, the Hamiltonian reads

$$\tilde{H} = \tilde{H}_e + \tilde{H}_{ph} + \tilde{H}_I + H_I' \tag{A.12}$$

where, taking β to be real,

$$\tilde{H}_e = H_e + \omega_0 \beta^2 + 2\beta M \tag{A.13}$$

$$\tilde{H}_I = [\omega_0 \beta + M](d + d^\dagger) = \tilde{M}(d + d^\dagger) \tag{A.14}$$

The system is initially in the state $|1;0> = |1>|0>$, where $|1>$ denotes the initial electronic state (usually, but not necessarily, the ground state) and $|0>$ denotes the vacuum for both the d and b$'$ photons.

A.3 Weak Second Harmonic Field

Let $|n>$ be the state of the field with n photons of type d and no b$'$ photons. If the b$'$ photon were in a resonating cavity, the probability that the system ends up with one b$'$ photon would be given by

$$P(v) = \lim_{t \to \infty} \sum_{n} \sum_{i=1}^{N} \left| <i;n| b' \exp[-i\tilde{H}t]|1;0> \right|^2 \tag{A.15}$$

This would contain complicated processes in which many such photons are emitted and absorbed many times. Here, we wish to exclude such processes on the grounds that the emitted photon does not interact again with the system: it is a running wave photon and, once it is emitted, it simply goes to the detector. We define the probability as the above expression with the proviso that H$'_I$ acts only once. Therefore, we set $\tilde{H} = H_0 + H'_I$, and write

$$P(v) = \lim_{t \to \infty} \sum_{n} \sum_{i=1}^{N} |A_i(n,t)|^2 \tag{A.16}$$

where

$$A_i(n,t) = -i\Theta(t) \int_0^t dt' <i;n| b' \exp[-iH_0(t-t')]H'_I \exp[-iH_0 t']|1;0> \tag{A.17}$$

The Fourier transformed amplitude

$$A_i(n,\omega) = \int_{-\infty}^{\infty} dt A_i(n,t) \exp[i\omega t]$$
$$= i <i;n| b'[\omega - H_0 + i\Gamma]^{-1} H'_I [\omega - H_0 + i\Gamma]^{-1}|1;0> \tag{A.18}$$

where Γ is a positive infinitesimal, may be expanded, by inserting a complete set, in the form

$$A_i(n,\omega) = i \sum_{m,k,p} \Psi_{ik}(n,m;\omega - \omega') M'_{kp} \Psi_{p1}(n,0;\omega) \tag{A.19}$$

where

$$\Psi_{ik}(n,m;\omega) = <i;n|[\omega - H_0 + i\Gamma]^{-1}|k;m>$$ (A.20)

These are the Excitation Amplitudes which can be computed exactly. This technique and its applications have been reviewed elsewhere [4].

A.4 Long-time Limit

We wish to obtain the asymptotic behaviour directly, without having to compute the Fourier transform; accordingly, we take the electron-phonon couplings to be infinitesimally damped. We must be careful since the rate of emitting photons of frequency v does not go to any limit as $t \to \infty$, but keeps oscillating forever. However, the measuring apparatus will average those oscillations over long times, which allows a limiting spectrum to be properly defined. Our final expression for the limiting rate is

$$R(v) = 2\Gamma \sum_{i,n,\lambda} \left| \sum_{m,k,p;\lambda'} \frac{R_{ik}^{(\lambda)}(n,m) M'_{kp} R_{p1}^{(\lambda')}(m,0)}{v + \eta\lambda - \eta\lambda' - i\Gamma} \right|^2$$ (A..21)

where $\eta\lambda$ and $R_{ik}^{(\lambda)}(n,m)$ are poles and residues of the Excitation Amplitudes:

$$\Psi_{ik}(n,m;\omega) = \sum_{\lambda} \frac{R_{ik}^{(\lambda)}(n,m)}{\omega - \eta\lambda + i\Gamma}$$ (A.22)

This completes the formalism.

References

1. Landau, L.D., Lifshitz, E.M.: Physique Statistique. Editions Mir, Moscou 1967
2. Bagchi, A., Barrera, R.G., Rajagopal, A.K.: Physical Review *B 20*, 4824 (1979)
3. Del Sole, R.: Solid State Communications *37*, 537 (1981)
4. Cini, M., D'Andrea, A.: Journal of Physics C: Condensed Matter *21*, 193 (1988)

Appendix B. SHG Slab Calculations

Michele Cini,[1] Rodolfo Del Sole,[1] and Lucia Reining[2]

[1] Dipartimento di Fisica, II Università di Roma, 'Tor Vergata', Via della Ricerca Scientifica, I-00133 Roma, Italia

[2] Centre Européen de Calcul Atomique et Moleculaire, Bât. 506, Université Paris Sud, Orsay, France

B.1 Theory

Let us consider a semi-infinite crystal, which is centrosymmetric in the bulk, occupying the half-space $z > 0$. Then, within the dipole approximation, any second-harmonic generation must necessarily come from the surface. We consider the case of radiation of frequency ω incident on the sample in the xz-plane at an angle ϕ_0 with respect to the normal to the surface, with wavevector component in the surface plane

$$Q(\omega) = (\omega/c)\sin\phi_0 \tag{B.1}$$

and
$$q_z(\omega) = (\omega/c)\cos\phi_0 \tag{B.2}$$

along z. Then the x-component of the vector potential of the emitted p-polarised light of frequency, 2ω, is given by (see Appendix A):

$$A_x(x,z;t) = [8\pi e^3 \alpha(2\omega)/\hbar m^3 c^3 S]$$
$$\times [1 - r_p^0(2\omega)^2]\exp\{i[Q(2\omega)x - q_z(2\omega)z - 2\omega t]\} \tag{B.3}$$
$$\times \sum_{j,k}[X_{xjk}(\omega) + \{Q(2\omega)/k_z(2\omega)\}X_{zjk}(\omega)]A_j^s A_k^s$$

where S is the surface area, $r_p^0(2\omega)$ is the reflection coefficient of p-polarised light,

$$\alpha(\omega) = (c/\omega)^2 [\varepsilon(\omega)q_z(\omega) + k_z(\omega)] / i4\varepsilon(\omega) \qquad (B.4)$$

and
$$k_z(\omega) = [(\omega/c)^2 \varepsilon(\omega) - Q^2]^{1/2} \qquad (B.5)$$

is the z-component of the light wave-vector in the crystal, j and k run over the cartesian coordinates x, y, z, and A^s is the vector potential of the incident light just below the surface. Equation B.3 has been derived from equations 6 and 19 of Ref. 1; according to this, we have assumed that the fields undergo a sudden transition from the vacuum values to bulk ones, namely the two media model of Fresnel optics. Later, we will generalize the results to the three-layer model, where an intermediate layer mimicking the surface is introduced. The z-component of the emitted vector potential in vacuum is easily found by exploiting the fact that A must be perpendicular to the wave vector:

$$A_z(x, z; t) = [Q(2\omega)/q_z(2\omega)]A_x(x, z; t) \qquad (B.6)$$

The key quantity in equation B.3 is $X_{ijk}(\omega)$, which is given by

$$
\begin{aligned}
X_{ijk}(\omega) = \sum_s f(s) \sum_r [1 - f(r)] \sum_n [1 - f(n)] \\
\times \{p_{sn}^i p_{nr}^j p_{rs}^k / [(E_{rs} - \hbar\omega - i0)(E_{ns} - 2\hbar\omega - i0)] \\
+ p_{sn}^i p_{nr}^j p_{rs}^k / [(E_{rs} - \hbar\omega - i0)(E_{ns} + \hbar\omega + i0)] \\
+ p_{sn}^i p_{nr}^j p_{rs}^k / [(E_{rs} + 2\hbar\omega + i0)(E_{ns} + \hbar\omega + i0)]\} \\
- \sum_s f(s) \sum_m f(m) \sum_n [1 - f(n)] \\
\times \{p_{mn}^i p_{sm}^j p_{ns}^k / [(E_{ns} - \hbar\omega - i0)(E_{nm} - 2\hbar\omega - i0)] \\
+ p_{mn}^i p_{sm}^j p_{ns}^k / [(E_{ns} - \hbar\omega - i0)(E_{nm} + \hbar\omega + i0)] \\
+ p_{mn}^i p_{sm}^j p_{ns}^k / [(E_{ns} + 2\hbar\omega + i0)(E_{nm} + \hbar\omega + i0)]\}
\end{aligned}
$$
$$(B.7)$$

where s, m, n, r label one-electron states, $f(s)$ is the occupation number of the s-state, 0 stands for a positive infinitesimal, and p_{sn}^i is the matrix element of the i-component of the momentum operator between states s and n. An alternative expression, more suitable for computation, can be derived from equation B.7 by simple algebraic manipulation:

$$X_{ijk}(\omega) = \sum_s f(s) \sum_r [1 - f(r)] \sum_n [1 - f(n)]$$

$$\times \{[p_{sn}^i p_{nr}^j p_{rs}^k / (E_{ns} - 2E_{rs})][1/(E_{rs} - \hbar\omega - i0) - 2/(E_{ns} - 2\hbar\omega - i0)]$$

$$+ [p_{sn}^i p_{nr}^j p_{rs}^k / (E_{ns} + E_{rs})][1/(E_{rs} - \hbar\omega - i0) + 1/(E_{ns} + \hbar\omega + i0)]$$

$$+ [p_{sn}^i p_{nr}^j p_{rs}^k / (E_{rs} - 2E_{ns})][1/(E_{ns} + \hbar\omega + i0) - 2/(E_{rs} + 2\hbar\omega + i0)]\}$$

$$- \sum_s f(s) \sum_m f(m) \sum_n [1 - f(n)]$$

$$\times \{[p_{mn}^i p_{sm}^j p_{ns}^k / (E_{nm} - 2E_{ns})][1/(E_{ns} - \hbar\omega - i0) - 2/(E_{nm} - 2\hbar\omega - i0)]$$

$$+ p_{mn}^i p_{sm}^j p_{ns}^k / (E_{nm} + E_{ns})][1/(E_{ns} - \hbar\omega - i0) + 1/(E_{nm} + \hbar\omega + i0)]$$

$$+ p_{mn}^i p_{sm}^j p_{ns}^k / (E_{ns} - 2E_{nm})][1/(E_{nm} + \hbar\omega + i0) - 2/(E_{ns} + 2\hbar\omega + i0)]\}$$

$$(B.7a)$$

It is straightforward to calculate the electric field from the emitted vector potential and, from this, the intensity $I^{2\omega}$ of the emitted SH light. We find:

$$I_p^{2\omega}/I_0^{\omega 2} = (512\pi^3 e^6/c^5) \left| \alpha(2\omega) \left[1 - r_p^0(2\omega)^2 \right] / \hbar\omega \cos\phi_0 \right|^2$$

$$\times \left| \sum_{j,k} \left[X_{xjk}(\omega) + \{Q(2\omega)/k_z(2\omega)\} X_{zjk}(\omega) \right] \left[E_j^s(\omega) E_k^s(\omega) / |E_0(\omega)|^2 \right] \right|^2$$

$$(B.8)$$

where I_0^ω and $E_0(\omega)$ are the intensity and the electric field of the incident light. The superscript p means that the emitted light is p-polarised. The factor $E_j^s(\omega) E_k^s(\omega)/|E_0(\omega)|^2$ in equation B.8, which relates the transmitted below-surface field, $E^s(\omega)$, to the incident one, is again found from Fresnel formulae; it also accounts for the polarisation of the incident field.

A similar calculation can be carried out for the intensity of the emitted s-polarised light, using equation 18, rather than equations 19 and 20 of Ref. 1; the result is:

$$I_s^{2\omega}/I_0^{\omega 2} = (512\pi^3 e^6/c^5) \left| \alpha(2\omega) \left[1 - r_p^0(2\omega)^2 \right] / \hbar\omega \right|^2$$

$$\times \left| \{\varepsilon(2\omega)\cos\phi_0 + [\varepsilon(2\omega) - \sin^2\phi_0]^{1/2}\} / \{\cos\phi_0 + [\varepsilon(2\omega) - \sin^2\phi_0]^{1/2}\} \right|^2$$

$$\times \left| \varepsilon(2\omega) - \sin^2\phi_0 \right|^{-1} \left| \sum_{j,k} X_{yjk}(\omega) \left[E_j^s(\omega) E_k^s(\omega) / |E_0(\omega)|^2 \right] \right|^2$$

$$(B.9)$$

Although the present theory avoids the explicit calculation of the non-local second-order susceptibility, nevertheless the spatial integral of the latter can be extracted from equation B.3. The structure of this equation suggests, indeed, that the 2ω vector potential far away from the surface is generated by a surface polarisation, $P_i(2\omega)$, proportional to $\sum_{jk} X_{ijk}(\omega)A_j^s A_k^s$. By comparing equation B.3 with the electric field generated by a surface polarisation $P(2\omega)$ localized in a depth, d, much smaller than the wavelength of light [2], we find

$$P_i(2\omega) = i4(c/\omega d)(e/mc)^3 \sum_{j,k} X_{ijk}(\omega)A_j^s A_k^s \tag{B.10}$$

from which the z-integral of the second-order suceptibility is obtained as:

$$\chi_{ijk}^{(2)}(\omega) \equiv \int_0^d dz \int_0^d dz' \int_0^d dz'' \chi_{ijk}^{(2)}(\omega, \omega; z, z', z'')$$
$$= -i4(e/m\omega)^3 X_{ijk}(\omega) \tag{B.11}$$

In writing equation B.11 we have used the fact that in centrosymmetric crystals the second-order suceptibility vanishes outside the surface layer.

The induced polarisation must be real, which implies that

$$X_{ijk}(-\omega) = X_{ijk}^*(\omega) \tag{B.12}$$

The real part of $X_{ijk}(\omega)$, proportional to the imaginary part of $\chi_{ijk}^s(\omega)$, can be calculated according to equation B.7 as

$$\begin{aligned}
X_{ijk}'(\omega) = &i\pi \sum_s f(s) \sum_r [1-f(r)] \sum_n [1-f(n)] \\
&\times \{[p_{sn}^i p_{nr}^j p_{rs}^k / (E_{ns} - 2E_{rs})][\delta(E_{rs} - \hbar\omega) - 2\delta(E_{ns} - 2\hbar\omega)] \\
&+ [p_{sn}^i p_{nr}^j p_{rs}^k / (E_{ns} + E_{rs})][\delta(E_{rs} - \hbar\omega) - \delta(E_{ns} + \hbar\omega)] \\
&+ [p_{sn}^i p_{nr}^j p_{rs}^k / (E_{rs} - 2E_{ns})][-\delta(E_{ns} + \hbar\omega) + 2\delta(E_{rs} + 2\hbar\omega)]\} \\
&- \sum_s f(s) \sum_m f(m) \sum_n [1-f(n)] \\
&\times \{[p_{mn}^i p_{sm}^j p_{ns}^k / (E_{nm} - 2E_{ns})][\delta(E_{ns} - \hbar\omega) - 2\delta(E_{nm} - 2\hbar\omega)] \\
&+ [p_{mn}^i p_{sm}^j p_{ns}^k / (E_{nm} + E_{ns})][\delta(E_{ns} - \hbar\omega) - \delta(E_{nm} + \hbar\omega)] \\
&+ [p_{mn}^i p_{sm}^j p_{ns}^k / (E_{ns} - 2E_{nm})][-\delta(E_{nm} + \hbar\omega) + 2\delta(E_{ns} + 2\hbar\omega)]\}
\end{aligned} \tag{B.13}$$

In spite of the imaginary unit at the beginning of equation B.13, this quantity is real, since the matrix elements of the momentum operator between real wave functions are purely imaginary (the wave functions can be taken to be real in any system without an external magnetic field). For this very reason, the quantity is also an even function of ω, in agreement with equation B.12. The imaginary part $X''_{ijk}(\omega)$ can be obtained, according to equation B.7a, by a Kramers-Krönig transform of equation B.13. Because $X'_{ijk}(\omega)$ is an even function of ω, the transform can be written as

$$X''_{ijk}(\omega) = (2\omega/\pi)\mathcal{P}\int_0^\infty d\omega' X'_{ijk}(\omega')/[\omega'^2 - \omega^2]$$
(B.14)

Since $\chi^{(2)}_{ijk}(\omega)$ is finite at zero frequency, because of equation B.11 $X_{ijk}(\omega)$ must vanish as, or faster than, ω^3 as ω goes to zero. This is clearly true for the real part, which is zero below half of the gap. In order that this also be true for the imaginary part, in view of equation B.14 we must have

$$\int_0^\infty d\omega' X'_{ijk}(\omega')/\omega'^2 = 0$$
(B.15)

which, because of equation B.11, is equivalent to

$$\int_0^\infty d\omega' \chi^{(2)''}_{ijk}(\omega')\omega' = 0$$
(B.16)

This is just the generalized f-sum rule derived by Scandolo and Bassani for nonlinear optics [3].

References

1. Cini, M. : Physical Review *B 43*, 4792 (1991)
2. Bagchi, A., Barrera, R.G., Rajagopal, A.K.: Physical Review *B 20*, 4824 (1979)
3. Scandolo, S., Bassani, F.: Physical Review *B 44*, 8466 (1991)

Appendix C. SHG Supercell Formalism

Raffaello Girlanda and Edoardo Piparo

Istituto di Struttura della Materia, Facoltà di Scienze m.f.n.,Università di Messina, Sant'Agata - Messina, Italia

C.1 Supercell Formalism

We expand the polarisation density, $\mathbf{P}(z,t)$, as a series of terms up to second order in the Fourier component of the field, obtaining the Fourier components of the polarisation:

$$P_\alpha^{(1)}(z;\omega') = \int_0^d dz' \chi_{\alpha\beta}^{(1)}(z;z',\omega')[E_\beta^{ex}(\omega') + E_\beta^{(1)}(z';\omega')] \tag{C.1}$$

$$\begin{aligned} P_\alpha^{(2)}(z;\omega') = &\int_0^d dz' \chi_{\alpha\beta}^{(1)}(z;z',\omega'+\omega'')E_\beta^{(2)}(z';\omega',\omega'') \\ &+ \int_0^d dz' \int_0^d dz'' \chi_{\alpha\beta\gamma}^{(2)}(z;z',\omega';z'',\omega'')[E_\beta^{ex}(\omega') + E_\beta^{(1)}(z';\omega')] \\ &\times [E_\gamma^{ex}(\omega'') + E_\gamma^{(1)}(z'';\omega'')] \end{aligned} \tag{C.2}$$

where $\chi^{(1)}$ and $\chi^{(2)}$ are the first and second order susceptibilities. The polarisation density is related to the density of the current, $\mathbf{J}$, by:

$$\frac{\partial}{\partial t}\mathbf{P}(z,t) = \mathbf{J}(z,t) - \mathbf{J}_{dc}(z) \tag{C.3}$$

where $\mathbf{J}_{dc}$ is the density of direct current.
The z-dependent density of the current for a given supercell is:

$$J(z,t) = -(ed/\Omega) \int_{cell} dr' \delta(z'-z) <r'|v\rho(t)|r'> \tag{C.4}$$

where Ω is the volume of the supercell.

If the density of direct current is negligible, one can obtain the following expressions for the bulk first susceptibility:

$$\chi_{\alpha\beta}^{(1)} = -d \frac{e^2}{\omega'^2} \int \frac{dk}{(2\pi)^3} \sum_{jj'JJ'} \frac{\Delta f_{jj'}^k}{\hbar\omega' - \Delta E_{j'j}^k} v_{\alpha Jj'}^k v_{\beta J'j}^k P_{zjJ}^k P_{zj'J'}^k \tag{C.5}$$

where $v_{\alpha Jj'}^k$ is the α-th cartesian component of

$$v_{Jj'}^k = <\psi_{Jk}|v| \psi_{j'k}> = \int_{cell} dr\, \psi_{Jk}^*(r) \frac{1}{i\hbar}[r,H]\psi_{j'k}(r) \tag{C.6}$$

f_{jk} is the occupation number of the state jk, and

$$P_{zjJ}^k = \int_{cell} dr'\, \psi_{jk}^*(r') \delta(z'-z)\, \psi_{jk}(r') \tag{C.7}$$

In order to determine the unknown induced electric field, E^{ind}, one has to solve, self-consistently, the following equation:

$$\nabla \times \nabla \times E^{ind}(z,t) + \frac{1}{c^2} \frac{\partial^2 E^{ind}(z,t)}{\partial t^2} = -\frac{4\pi}{c^2} \frac{\partial^2 P(z,t)}{\partial t^2} \tag{C.8}$$

which is obtained by eliminating the magnetic field from the Maxwell equations. It is convenient to introduce a set of Green functions, $G_\alpha(z,z';\omega)$, satisfying the following set of equations and boundary conditions:

$$(1-\delta_{\alpha z}) \frac{1}{q^2} \frac{\partial^2 G_\alpha(z,z';\omega)}{\partial z^2} + G_\alpha(z,z';\omega) = \delta(z-z') \tag{C.9}$$

$$G_\alpha(0,z';\omega) = G_\alpha(d,z';\omega) \tag{C.10}$$

$$\left.\frac{\partial G_\alpha(z,z';\omega)}{\partial z}\right|_{z=0} = \left.\frac{\partial G_\alpha(z,z';\omega)}{\partial z}\right|_{z=d} \tag{C.11}$$

where $q = (\omega/c)$. It is then easily demonstrated that:

$$G_z(z, z'; \omega) = \delta(z - z') \qquad (C.12)$$

while for $\alpha \neq z$ one has:

$$G_\alpha(z, z'; \omega) = \frac{q}{2\sin(qd/2)}\cos\left[q(|z - z'| - \tfrac{1}{2}d)\right] = \frac{1}{d} + O(q^2) \qquad (C.13)$$

Thus:

$$E_\alpha^{(1)}(z; \omega') = -4\pi \int_0^d dz'\, G_\alpha(z, z'; \omega') P_\alpha^{(1)}(z'; \omega') \qquad (C.14)$$

Introducing expression C.14 into C.1, and making use of equation 7.11, one obtains a set of integral equations for $\mathbf{P}^{(1)}(z; \omega')$ that can be iterated obtaining:

$$P_\alpha^{(1)}(z; \omega') = \tilde{\chi}_{\alpha\beta}^{(1)}(z; \omega') E_\beta^{ex}(\omega') \qquad (C.15)$$

where the *effective* susceptibility, $\tilde{\chi}^{(1)}$, is given by:

$$\tilde{\chi}_{\alpha\beta}^{(1)}(z; \omega') = \int_0^d dz'\, \chi_{\alpha\beta}^{(1)}(z; z', \omega') - 4\pi \int_0^d dz'\, \chi_{\alpha\beta'}^{(1)}(z; z', \omega') \int_0^d dz_1 G_{\beta'}(z', z_1; \omega')$$

$$\times \int_0^d dz_2 \chi_{\beta'\beta}^{(1)}(z_1; z_2, \omega') \;+ \dots$$

$$(C.16)$$

A certain simplification of the above formulas can be obtained neglecting the terms of order higher than q in equation C.13.

Subject Index

Esprit Basic Research Series

J. W. Lloyd (Ed.): **Computational Logic.** Symposium Proceedings, Brussels, November 1990. XI, 211 pages. 1990

E. Klein, F. Veltman (Eds.): **Natural Language and Speech.** Symposium Proceedings, Brussels, November 1991. VIII, 192 pages. 1991

G. Gambosi, M. Scholl, H.-W. Six (Eds.): **Geographic Database Management System.** Workshop Proceedings, Capri, May 1991. XII, 320 pages. 1992

R. Kassing (Ed.): **Scanning Microscopy.** Symposium Proceedings, Wetzlar, October 1990. X, 207 pages. 1992

G. A. Orban, H.-H. Nagel (Eds.): **Artificial and Biological Vision Systems.** XII, 389 pages. 1992

S. D. Smith, R. F. Neale (Eds.): **Optical Information Technology.** State-of-the-Art Report. XIV, 369 pages. 1993

Ph. Lalanne, P. Chavel (Eds.): **Perspectives for Parallel Optical Interconnects.** XIV, 417 pages. 1993

D. Vernon (Ed.): **Computer Vision: Craft, Engineering, and Science.** Workshop Proceedings, Killarney, September 1991. XIII, 96 pages. 1994

E. Montseny, J. Frau (Eds.): **Computer Vision: Specialized Processors for Real-Time Image Analysis.** Workshop Proceedings, Barcelona, September 1991. X, 216 pages. 1994

J. L. Crowley, H. I. Christensen (Eds.): **Vision as Process.** Basic Research on Computer Vision Systems. VIII, 432 pages. 1995

B. Randell, J.-C. Laprie, H. Kopetz, B. Littlewood (Eds.): **Predictably Dependable Computing Systems.** XIX, 588 pages. 1995

F. Baccelli, A. Jean-Marie, I. Mitrani (Eds.): **Quantitative Methods in Parallel Systems.** XVIII, 298 pages. 1995

J. F. McGilp, D. Weaire, C. H. Patterson (Eds.): **Epioptics.** Linear and Nonlinear Optical Spectroscopy of Surfaces and Interfaces. XII, 230 pages. 1995

Springer-Verlag
and the Environment

We at Springer-Verlag firmly believe that an international science publisher has a special obligation to the environment, and our corporate policies consistently reflect this conviction.

We also expect our business partners – paper mills, printers, packaging manufacturers, etc. – to commit themselves to using environmentally friendly materials and production processes.

The paper in this book is made from low- or no-chlorine pulp and is acid free, in conformance with international standards for paper permanency.